JOURNAL OF SOIL CONTAMINATION

Volume 6 / Number 6

Editor

James Dragun, Ph.D.
Dragun Corporation
Farmington Hills, Michigan

Associate Editor
Paul Kostecki, Ph.D.
Northeast Environmental Public Health Center
School of Public Health
University of Massachusetts
Amherst, MA

Managing Editor
Barbara Knowles
Association for the
Environmental Health of Soils
Amherst, Massachusetts

CRC LEWIS PUBLISHERS

Boca Raton   New York

Chris Richardson: Journals Director
Jim McGovern: Senior Project Editor
Anita Klein: Senior Typesetter
Vikki Mitchell: Editorial Assistant

Journal of Soil Contamination is published bi-monthly by CRC Press LLC, 2000 Corporate Blvd., N.W., Boca Raton, FL 33431 USA. For 6 issues, U.S. rates are $430.00 to institutions; Society rate: $110.00 U.S.; add $7.95 per issue foreign shipping and handling fees to all orders not shipped to a United States or Canada zip code. For immediate service and charge card sales, call our toll-free number: **1-800-272-7737** Monday through Friday, 7:30 a.m.–6:00 p.m. EST (Continental U.S. and Canada only). To order by FAX: **1-800-374-3401**. Or send written orders to: **CRC PRESS LLC, Dept. 100, PO Box 6123, Fort Lauderdale, FL 33310.**
POSTMASTER: Send address changes to CRC Press LLC, 2000 Corporate Blvd., N.W., Boca Raton, Fl 33431.

ISSN 1058-8337

November 1997

©1997 by AEHS
Amherst

No claim to original U.S. Government works

Boca Raton   New York   London   Tokyo
Printed on acid-free paper

The *Journal of Soil Contamination* is abstracted in Cambridge Scientific Abstracts; abstracted and indexed in Chemical Abstracts Service and Environment Abstracts; and indexed in Environmental Periodicals Bibliography.

JOURNAL OF SOIL CONTAMINATION

CONTENTS

Volume 6 / Number 6

Aim and Scope

Journal of Soil Contamination is a new journal that will be the main vehicle of communication of the Association for the Environmental Health of Soils (AEHS). It will provide a direct link between the association's membership and those disciplines concerned with the technical, regulatory, and legal challenges of contaminated soils. The journal will be a bi-monthly, internationally peer-reviewed publication focusing on scientific and technical information, data, and critical analysis in the following areas:

- **Analytical chemistry,** including basic analytical problems with soil, product identification, development and evaluation of field screening and analytical techniques, and development and evaluation of laboratory analytical techniques and standards

- **Site assessment,** including field sampling techniques and statistical design, sample handling and preparation, and assessment methodologies

- **Environmental fate,** including the chemistry and physics influencing the movement and partitioning of contaminants in the soil

- **Environmental modeling,** including the mathematical representation of contaminant movement and its relationship to real world utility

- **Remediation techniques,** including both offsite and *in situ* techniques, such as vacuum extraction, bioremediation, low and high thermal treatment, land treatment, solidification and encapsulation, asphalt incorporation, chemical treatment, soil washing, and vitrification

- **Risk assessment issues,** including health effects and hazards, exposure assessment, and risk characterization of sites and actions (remediation)

- **Risk management** or the application of risk assessment information, especially during remediation

- **Regulatory issues,** especially numerical standards of cleanups (action levels, maximum concentration levels, and cleanup levels) and approaches/methodologies utilized by the regulatory community

- **Legal considerations** pertaining to regulatory statutes or actions, as well as those pertaining to the private sector, such as banking and real estate transactions

Journal of Soil Contamination will consider all types of soil contamination. Sludges and petroleum contamination and their chemical constituents are recent concerns; however, the journal will also discuss petrochemical, chlorinated hydrocarbon, pesticide and heavy metal (especially lead) contamination.

EDITOR'S NOTE

Why do we have journals? Are they really essential? The obvious answer is that journals are essential, because they disseminate information. However, equally important is that journals monitor and control the quality of published works in a discipline. The scientific community believes that it is beneficial for carefully responsible, chosen editors to monitor papers, and for anonymous reviewers to critique these papers and recommend changes when needed. This author-reviewer-editor relationship establishes the standard of quality of a journal. The standard of quality should be high, because it effects the overall quality and vitality of not only a scientific discipline, but, in the case of the *Journal of Soil Contamination,* the quality and vitality of our lives.

The environmental "movement" and our environmental regulatory system is a product of a process that began in earnest only about 3 decades ago. That process has yielded protection for the environment and for humans, protection that did not exist beforehand. Protection today and in the future depends on the quality of the data and information on which the regulatory system is based. A good regulatory system is based on knowledge revealed through published scientific papers that contain sound data and information.

As in the past, the *Journal of Soil Contamination* is committed to being the journal of record for important new results in the soil contamination arena. We will strive to maintain the highest standards of quality and innovation in the field. We realize that we have an important responsibility to produce the very best journal that can be published. In this age of information overload where scientists and engineers are being bombarded with more and more printed material, we are determined in 1997 to provide you with only the best, so that the *Journal of Soil Contamination* will always be required reading.

As we enter our sixth year, the editorial office of the *Journal of Soil Contamination* thanks its supporters and colleagues who have contributed to the success of this journal. We thank the many scientists who have submitted their papers for review. These dedicated individuals have devoted innumerable hours, many times without compensation and acknowledgment, for the sake of the advancement of science. We also thank our tireless and meticulous reviewers, who do their best to conduct reviews thoroughly and fairly. Peer review is the foundation on which science is based; without it and reviewers, the scientific literature would be awash with bad science and badly composed papers. Finally, we wholeheartedly thank our resourceful and loyal staff who handle so well those never-ending tasks that must be hurdled in order to publish this journal.

The *Journal of Soil Contamination* looks forward to becoming a greater force in 1997 in the scientific community, with the continued help and support of our subscribers, authors, reviewers, editorial board members, and our staff.

Every year in the U.S., several hundreds of millions of private industry and public taxpayer dollars are spent sampling, analyzing, delineating, excavating, hauling, capping, incinerating, or otherwise treating soils thought to pose a threat to the public health. For many soil contaminants, the investigation/risk assessment/ remediation process has become fairly streamlined and almost routine: standard analytical methods exist, the fate and transport characteristics and potential health effects are well understood, an array of proven remedial technologies are available, etc. Indeed, for some chemicals such as petroleum hydrocarbons, regulatory guidance documents describe a generic, step-by-step approach for addressing the impacted soils.

Chromium is not one of these chemicals. Indeed, when chromium is the primary soil contaminant, virtually every aspect of the environmental investigation process involves an unsettled issue or some measure of controversy. For example, in the past ten years, the EPA has issued, withdrawn and re-proposed promulgated methods for the analysis of hexavalent chromium [$Cr(VI)$] in soil; the EPA has also proposed and withdrawn a noncancer inhalation toxicity criterion for $Cr(VI)$ (no value currently exists) and is re-evaluating the cancer slope factor for $Cr(VI)$ based on new epidemiological data. Environmental levels of airborne $Cr(VI)$ (classified by EPA as a known inhalation carcinogen) could not be reliably measured until four years ago (with the development of a new analytical technique), and even today "background" levels of ambient $Cr(VI)$ have been measured only in certain parts of the U.S. Many of the outstanding issues have been vigorously disputed. For example, over thirty papers have been published in the last five years debating topics such as: can $Cr(III)$ in soil can plausibly be expected to oxidize to $Cr(VI)$ under environmental conditions? will dermal contact with $Cr(VI)$ in soil elicit allergic contact dermatitis? is urinary chromium sampling a reliable measure of exposure to chromium in household dusts? In short, the risk management of chromium-contaminated soils continues to be a very dynamic process that presents several interesting challenges.

Because of the many unique commercial properties of chromium, there have been instances in the past where manufacture or use of large quantities of chromium has resulted in elevated chromium levels in soils or groundwater. For example, because of the chromite-ore processing activities that occurred earlier in the century, approximately 3 million tons of chromium-containing fill are currently present at more than 200 properties in Hudson County, NJ. The historic use of $Cr(VI)$ as a corrosion inhibitor in cooling towers has resulted in $Cr(VI)$ groundwater plumes in several parts of the U.S. Many of the uncertainties identified above have surfaced during examination of these and other sites. The final outcome of these investigations and assessments (*e.g.*, the derived cleanup standard, the volume of material to be remediated, the remedial option of choice, claims of property damage) will ultimately drive the cost (public and private) and as a result, it is critical to ensure the investigation process be as free as possible of uncertainty and guesswork.

Several individuals in regulatory, academic, and industry settings are currently conducting research to bring some resolution or closure to these issues. At the 11th Annual Contaminated Soils Conference organized by the Association for the Environmental Health of Soils, ten of these scientists, including experts in toxicology, industrial medicine, risk assessment, chemistry, engineering and geology, presented their most recent thoughts and findings at a symposium entitled "Chromium in the Environment". The chapters of this volume, which are authored by the speakers, address numerous ground-breaking developments in various fields of environmental chromium research, including:

- the chromium chemistry necessary for understanding the fate and toxicity of chromium in the environment;

- analytical methods for Cr(VI) in soil;

- the unique conditions required for environmental oxidation of Cr(III) to Cr(VI);

- the use of human volunteers in chromium exposure studies designed to reduce uncertainty in risk assessment;

- approaches for setting health-based cleanup standards for chromium in soil and the basis and source of variability in state and federal standards;

- methods for managing impacts of Cr(VI)-containing residues on groundwater;

- dosimetry for cleanup levels based on allergic contact dermatitis;

- effects of chromium-containing soils on concrete foundations and structures; and

- technologies for remediating Cr(VI) in soil.

The material presented in these chapters should significantly advance our ability to assess and manage the potential health hazards posed by environmental chromium. While this volume focuses on a single element in a single medium, the applied research-oriented contents will hopefully stimulate scientists in related fields to identify problem-solving techniques for reducing uncertainty in their own fields of endeavor. A similar application of first principles and scientific method should allow us to eventually answer some of the more complex and pressing environmental issues of our day (*e.g.*, are estrogenic effects occurring in humans as a result of exposure to environmental chemicals? how large is the margin between "no-effect levels" posed by dioxin in humans and "effect levels" in animals, and what is the best dosimetric? are there quantifiable health effects from secondary tobacco smoke?).

My colleagues and I look forward to continued support at future symposia. My grateful thanks to the conference directors for their help, insight, and participation.

Brent L. Finley, Ph.D., D.A.B.T.

Journal of Soil Contamination, 6(6):561–568 (1997)

Chromium Chemistry and Implications for Environmental Fate and Toxicity

Joel Barnhart

American Chrome & Chemicals, 3900 Buddy Lawrence Drive, Corpus Christi, TX 78469

The same chemical properties that make chromium such an important component of so many industrial and consumer products are the important factors controlling its environmental fate and toxicity. Although only about 15% of the chromium mined is used in the manufacture of chromium chemicals, the chemistry of chromium is important in nearly all applications. For instance, the "stainless" nature of stainless steel is due to the chemical properties of the chromium oxides that form on the surface of the alloy. Similarly, the product protection afforded by chrome plating of metals, CCA treatment of wood, and chrome tanning of leather is directly dependent on chromium chemistry. In all of these applications the most important chemical property is that under typical environmental and biological conditions of pH and oxidation-reduction potential, the most stable form of chromium is the trivalent oxide. This form has very low solubility and low reactivity resulting in low mobility in the environment and low toxicity in living organisms. The chemical properties of the major commercial products of chromium are discussed in the context of the Eh-pH diagram. These same chemical properties control the environmental fate of chromium and are closely tied to the toxicity of the various compounds.

KEY WORDS: *chromium chemistry, stainless steel, chromium oxides.*

INTRODUCTION

*C*HROMIUM is one of the most widely used metals, yet in many ways it is misunderstood. It is used extensively because, in many forms, its stability helps protect materials from degradation by the environment. However, it is the subject of this issue because it can exist in forms that are toxic and can be hazardous to the environment. The following is an overview of the principal chromium chemicals and how they are related by their chemistry.

DISCUSSION

Chromium is the 24th element on the periodic chart. It is situated between vanadium and manganese and has an average atomic weight of 52. Chromium is the 21st most abundant element in the Earth's crust at about 100 ppm. The average soil in the U.S. contains about 40 ppm but ranges from less than 1 to greater than 1000 (USGS, 1984).

On a world-wide basis, about 80% of the mined chromium goes into metallurgical applications. Much of this goes into the manufacture of stainless steel. About 15% is used in the manufacture of chromium chemicals, and the remainder is used in refractory applications. For nearly all of these uses, the chemical properties of chromium are integral to its effectiveness. In metallurgical applications, the physical properties that chromium imparts to alloys are important, but it is the chemical properties derived from chromium that are essential in many of these uses. Indeed, chromium is what makes stainless steel "stainless". In refractory applications, the inert nature of trivalent chromium oxide, either by itself or in combination with other refractory oxides, such as those of iron, aluminum, and magnesium, is the reason it is used in the most corrosive environments.

The total amount of chromium chemicals currently used annually in the U.S. is similar to that used in the early 1950s. Based on the authors knowledge of the current market for chromium chemicals and information contained in contemporary reports from the earlier period (Copson, 1956), it is estimated that about 110,000 tons of sodium dichromate dihydrate equivalent will be used in 1996 vs. about 120,000 tons in 1951. Some of this lack of growth in the market is due to replacement of chromium by other materials, as in pigments, or the transfer of manufacturing operations overseas, as in leather tanning. However, in several cases better utilization is also an important factor. Table 1 gives the approximate distribution of the use of chromium chemicals in the major applications in the U.S. and the Western world in 1996 with a comparison to use in the U.S. in 1951. During this period the only significant increase in the use of chromium chemicals in the U.S. has come in wood-treating applications.

TABLE 1
Chromium Chemicals: Use by Industry

	1996 Western world	1996 USA	1951 USA
Wood preservation	15%	52%	2%
Leather tanning	40%	13%	20%
Metal finishing	17%	13%	25%
Pigments	15%	12%	35%
Refractory	3%	3%	1%
Other	10%	7%	17%

Note: Adapted from Barnhart (1997).

Three chemical characteristics of chromium, which are fundamental to understanding its many uses as well as its impact on human health and the environment, can be summarized in the following statements (Barnhart, 1997).

1. The dominant naturally occurring form of chromium is the trivalent oxide.

2. Other forms will tend to be converted to the trivalent oxide when in contact with the natural environment.

3. Even when put in environments where it is not thermodynamically stable, trivalent chromium oxide is very slow to react.

The first of these is confirmed by the scarcity of reports of naturally occurring chromium existing in any other oxidation state (Goldschmidt, 1958). Chromium can exist in every oxidation state from −2 to +6 but only the trivalent iron chromite is found in large enough deposits to be commercially mined (Sulley and Brandes, 1967). Chromite ore has a spinel crystal structure containing trivalent chromium, iron, and small amounts of other divalent and trivalent metal oxides such as aluminum and magnesium (Udy, 1956). This is a very stable structure and consequently is useful in high-temperature applications. Similar chromium compounds are found in most soils. The few instances where hexavalent chromium has been found to be naturally occurring are in locations with very unusual local environments (Robertson, 1975).

The second point is really the mechanism that leads to the first, but it is so important that it is listed separately. The trivalent oxidation state is thermodynamically the preferred form in most natural environments (Bartlett and James, 1988; James, 1996). The significance of this is emphasized later when the chemistry of the major commercial applications of chromium are discussed.

The third point describes an important kinetic property of chromium oxide reactions that is significant in many applications (Cotton and Wilkinson, 1980).

A key factor in understanding chromium chemistry is understanding the conversions from one oxidation state to another. The energy associated with an element changing oxidation state is often described in terms of the reduction potentials of the species involved. While this information is vital to understanding the chemistry, it is somewhat difficult to relate to real world situations. Because reduction potentials are normally given for species in their standard states, they must be corrected for concentration and especially pH to use them in most applications. These mathematical manipulations, although well documented, are often tedious to perform.

A pictorial way to illustrate these important chemical properties is by using an Eh-pH diagram. This technique is widely used in geochemistry and the study of corrosion but is not as common in other fields. The most complete compilation of Eh-pH diagrams is by Marcel Pourbaix (1974). The ones presented here are adapted from Pourbaix but are somewhat simplified to emphasize the main points. The species indicated on the diagrams are meant to represent classes of commercial chemicals rather than the specific nature of the ions in solution. Nonetheless, the thermodynamic principles remain the same.

By using the Eh-pH diagram, many important chemical properties of chromium can be illustrated. Because Eh-pH diagrams describe reactions in aqueous solution, the diagram for water is fundamental. The usual convention of showing the stability region for water with dotted lines is used here. Above the upper line, water can be oxidized to oxygen molecules and hydrogen ions. Below the bottom line, water can be reduced to hydrogen molecules and hydroxyl ions. A reminder of this is shown on the right side of each of the diagrams.

In considering the impact of chromium on human health and the environment, we must consider its interaction with various biological systems. Some of the most fundamental chemical reactions in living organisms are those that convert carbon dioxide to organic molecules and organic molecules to carbon dioxide. These reactions and many other oxidation-reduction reactions involving organic matter have standard reduction potentials near that of water reduction to hydrogen. As most living systems operate somewhere near neutral pH, there is a limited region on the Eh-pH diagram where much of the chemistry of biological systems takes place. This location is important in understanding the reactions of chromium in biological systems because a substance that is thermodynamically stable in this region is unlikely to react in biological systems.

In Figure 1, the reactions that produce the major chromium chemicals are summarized. The first reaction considers the conversion of chromite ore to sodium chromate. This is the fundamental reaction for producing the basic chromium chemicals. This reaction is thermodynamically possible in aqueous solution with the right oxidant as both species are stable in water. However, no process based on the aqueous oxidation of chromite ore has been commercialized. This is a direct result of the third fundamental point of chromium chemistry listed above. Even when put in environments where it is not thermodynamically stable, trivalent chromium oxide is very slow to react. Commercially, sodium chromate is made by reacting the ore with an alkaline material such as sodium carbonate or sodium

CHROMIUM CHEMICALS MANUFACTURE

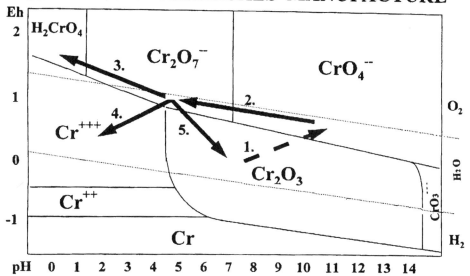

FIGURE 1

*Basic chromium chemicals manufacture. (1) Sodium chromate; (2) sodium dichromate;
(3) chromic acid; (4) basic chromium sulfate; (5) chromic oxide (Barnhart, 1997).*

hydroxide at temperatures above 1000°C in an excess of oxygen (Hartford, 1979). This is a much more difficult process to operate than an aqueous reaction but is necessary because of the low solubility and reactivity of the trivalent chromium oxide in the ore. Because this process does not occur in solution, it is indicated on the diagram with a dashed arrow.

To produce other chromium chemicals from sodium chromate, the Eh and/or pH of the system must be changed. These changes are indicated by solid arrows in Figure 1. Sodium dichromate and chromic acid are produced by lowering the pH. This is usually done with sulfuric acid, but carbon dioxide and electrochemical generation of hydrogen ions are also used commercially for the acidification steps (Copson, 1956; Barnhart, 1990). The trivalent basic chrome sulfate is produced by lowering both the Eh and pH of a sodium dichromate solution. An organic material like sugar or an inorganic reductant like sulfur dioxide is used to lower the Eh, and sulfuric acid is usually used to lower the pH (Copson, 1956).

Eh-pH diagrams can also be used to describe the chemical reactions of chromium in the major applications. These are shown in Figure 2. In addition, the approximate location of the region where most biological activity takes place is indicated by the ellipse. The first process considered is wood treatment. Either sodium dichromate or chromic acid can be used, but in the U.S. it is usually

PRINCIPAL USES OF CHROMIUM

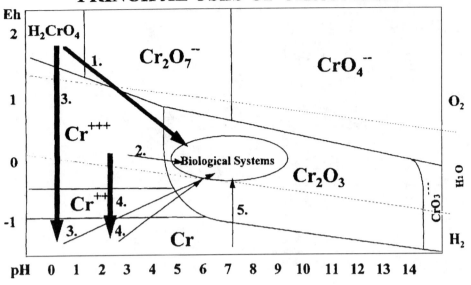

FIGURE 2

*Principal uses of chromium and their relationship to biological systems. (1) Wood treatment;
(2) leather tanning; (3) chromic acid plating; (4) trivalent chromium plating; (5) stainless steel
(Barnhart, 1997).*

chromic acid (Arsenault, 1980). Chromic acid is combined with chemicals like
copper oxide and arsenic acid, which are toxic to the organisms that decompose
wood. The resulting solution is then forced into the wood under pressure. Once
inside, the hexavalent chromium is reduced to the trivalent form by organic
compounds in the wood and becomes insoluble. This process "fixes" the copper
and arsenic in the wood along with the chromium. Thus, the wood will be resistant
to decay even in wet environments for 40 years or more because the chromium,
copper, and arsenic remain in place.

The next application, leather tanning, has similarities with wood treatment. In
fact, some of the earliest processes for chrome tanning of leather used hexavalent
chromium chemicals to saturate the skin and then reduced them to insoluble forms
in place. Now, however, the standard practice is to use a soluble trivalent compound and a "masking agent", which allows the chromium to effectively penetrate
the hide (Lollar, 1956; Lollar, 1980). In this case much of the chromium appears
to complex with proteins in the leather, which greatly slows down the final
transformation to the chromic oxide form. Because the chromium in this application is also "fixed", it imparts to leather the water resistance and flexibility over
extended periods of time that are essential in most of its applications.

The most familiar use of chromium is chrome plating. By putting a thin layer of chromium on an object, the lifetime can be greatly extended and the appearance enhanced. There are two broad categories of plating, functional, and decorative (Sully and Brandes, 1967). In functional plating, the chromium surface is usually put there to improve the wear resistance. Piston rings and crankshafts are examples. In decorative plating, appearance and corrosion resistance are the important properties, and the deposits are usually much thinner. Essentially all functional plating uses chromic acid solutions, while decorative plating may be done with either a chromic acid or a soluble trivalent chromium solution. The arrows for these two types of plating illustrate the chemical reactions involved. Both are done in an acidic environment using an electrical potential to perform the chemical reduction. As either a three- or six-electron change is involved, a substantial current is required even though relatively thin deposits are produced. Also, a catalyst is required to achieve reasonable current efficiencies. The chemical reaction for these industrial processes are shown with the heavy arrow. The second reaction, shown with a thin arrow, indicates the formation of trivalent chromium oxide on the surface. This is often the vital step in making the application successful.

From the earlier discussions, it may seem surprising that chromium metal, which is not thermodynamically stable in the stability region of water, can stay bright and shinny on a kitchen faucet and not be easily oxidized. In fact, the metal remains intact because, like several other metals, a protective layer of oxide forms on its surface. This thin, invisible film of trivalent chromium oxide is so unreactive and impenetrable that under most conditions it protects the underlying metal from being exposed to the environment and rapidly corroding. Thus, it is really the chemical compound chromium oxide that is in contact with the environment.

A very similar mechanism gives stainless steel its stainless character (Wallen and Olsson, 1977). Many of the uses of stainless steel depend on the physical properties that chromium imparts to steel, such as hardness, strength, and high-temperature resistance. However, if the chromium metal continuously reacted with oxygen in moist environments, it would seldom be used. Again in the case of stainless steel, it is the inert nature of the chromium oxide that forms on the surface that is essential for the many valuable applications. This surface reaction is indicated by a thin arrow. In other alloys containing chromium the same chemical properties are often vital.

In refractory applications of chromium, the stability of the trivalent oxide is fundamental (Heuer *et al.*, 1956). Chromium refractories are used in the most severe environments because of their ability to withstand the high temperatures and corrosive conditions present. In fact, a major factor in the decline in use of some chromium-containing refractories is not that better refractories have been found but instead that process changes have reduced the demand put on the refractories so that less-resistant materials can be used.

CONCLUSION

The three key points of chromium chemistry that were emphasized are the controlling factors in where chromium is found in nature, in how chromium is used commercially, and in how it behaves in most environments. Also, by using the Eh-pH diagram and remembering that chromic oxide reacts very slowly, if at all, in aqueous solution much of the chemistry of chromium can be readily understood.

REFERENCES

Arsenault, R. D. 1980. Chromium use in wood preservation, *Chromates Symposium 80 — Focus Of A Standard.* Industrial Health Foundation, Inc., Pittsburgh, pp. 224–226.

Barnhart, R. J. 1990. Chromic acid production, *AESF Second Chromium Colloquium.* American Electroplaters and Surface Finishers Society, Orlando.

Barnhart, R. J. 1997. Occurrences, uses, and properties of chromium. *Regulatory Toxicology and Pharmacology* **26,** 53–57.

Bartlett, R. J. and James, B. R. 1988. Mobility and bioavailability of chromium in soils. In: *Chromium in Natural and Human Environments*; pp. 267–304. (Nriagu, J. O. and Nieboer, E., Eds.) New York, Wiley-Interscience.

Copson, R. L. 1956. Production of chromium chemicals. In: *Chromium*; pp. 262–282. (Udy, M. J., Ed.) New York, Reinhold.

Cotton, F. A. and Wilkinson, G. 1980. *Advanced Inorganic Chemistry*, pp. 719–749. New York, John Wiley & Sons.

Goldschmidt, V. M. 1958. *Geochemistry.* London, Oxford University Press.

Hartford, W. H. 1979. Chromium compounds. In: *Encyclopedia of Chemical Technology*, pp.82–120. New York, John Wiley & Sons.

Heuer, R. P., Trostel, L. J., and Grigsby, C. E. 1956. Chromium in refractories. In: *Chromium*; pp. 327–390. (Udy, M. J., Ed.) New York, Reinhold.

James, B. R. 1996. The challenge of remediating chromium-contaminated soil. *Environ. Sci. Technol.* **30(6),** 248A–251A.

Lollar, R. M. 1956. Chromium chemicals in the tanning industry. In: *Chromium*; pp. 302–314. (Udy, M. J., Ed.) New York, Reinhold.

Lollar, R. M. 1980. Chromium use in the tanning industry, *Chromates Symposium 80— Focus Of A Standard.* Industrial Health Foundation, Inc., Pittsburgh, pp. 237–242.

Pourbaix, M. 1974. *Atlas of Electrochemical Equilibria in Aqueous Solutions.* National Association of Corrosion Engineers, Houston, pp. 256–271.

Robertson, F. N. 1975. Hexavalent chromium in the ground water in Paradise Valley, Arizona. *Ground Water.* **13(6),** pp. 516–527.

Sully, A. H. and Brandes, E. A. 1967. *Chromium*, 2nd ed. New York, Plenum Press.

Udy, M. J. 1956. Recovery of chromium from its ores. In: *Chromium*; pp. 3–4. (Udy, M. J., Ed.) New York, Reinhold.

USGS (United States Geological Survey). 1984. *Element Concentrations in Soils and Other Surficial Materials of the Conterminous United States.* USGS Professional Paper 1270. Washington, DC. U.S. Government Printing Office.

Wallen, B and Olsson, J. 1977. Corrosion resistance in aqueous media. In: *Handbook of Stainless Steels*; pp. 16–1–89. (Peckner, D. and Bernstein, I. M., Eds.) New York, McGraw-Hill.

Journal of Soil Contamination, 6(6):569–580 (1997)

Oxidation-Reduction Chemistry of Chromium: Relevance to the Regulation and Remediation of Chromate-Contaminated Soils

Bruce R. James,[1,] John C. Petura,[2] Rock J. Vitale,[3] and George R. Mussoline[3]*

[1]*Soil Chemistry Laboratory, College of Agriculture and Natural Resources, University of Maryland, College Park, MD;* [2]*Applied Environmental Management, Inc. Malvern, PA;* [3]*Environmental Standards, Inc., Valley Forge, PA*

Developing health-protective clean-up standards and remediation strategies for chromium-contaminated soils based on the hexavalent forms of this heavy metal is a complex and controversial issue, because certain forms of Cr(III) can oxidize to Cr(VI) and Cr(VI) can be reduced to Cr(III) under diverse soil conditions. The extent of oxidation of Cr(III) in soils amended with wastes is based on four interacting parameters:

(1) solubility and form of Cr(III) related to oxidation (waste oxidation potential, or WOP), (2) reactive soil Mn(III,IV) (hydr)oxide levels (soil oxidation potential for Cr(III), or SOP), (3) soil potential for Cr(VI) reduction (soil reduction potential, or SRP), and (4) soil-waste pH as a modifier of the first three parameters (pH modification value, or PMV). Each of these four parameters can be quantified with laboratory tests and ranked numerically; the sum of which is the Potential Chromium Oxidation Score (PCOS) for assessing the relative hazard of a waste-soil combination. The PCOS value is proposed as the basis for setting acceptable maximum limits for Cr(III)-containing wastes in particular soil environments to maintain Cr(VI) levels at or below health-based limits. Adjusting and controlling the PCOS parameters will allow practitioners to design remediation strategies and regulators to establish Cr clean-up standards under soil conditions that minimize Cr(III) oxidation and maximize Cr(VI) reduction. Effective remediation of Cr(VI)-contaminated soils by reduction depends on three principles: (1) reduction of Cr(VI) to forms of Cr(III) that are inert toward re-oxidation, (2) the absence of undesirable reaction products (e.g., the oxidized forms of certain reducing agents), and (3) establishment or maintenance of soil pH and Eh conditions that favor reduction of Cr(VI) and disfavor the oxidation of Cr(III). This article focuses on the importance of understanding the waste forms of Cr and the redox-related properties of Cr-contaminated soils in the design of remedial goals to protect human health and ecological function.

---

* Corresponding Author.

1058-8337/97/$.50

$\mathcal{H}$ EXAVALENT Cr exists in soils as a relatively soluble anion under most conditions (CrO_4^{2-} or $HCrO_4^-$); it is a Class A human carcinogen by inhalation and may be acutely toxic or irritating to living cells (Yassi and Nieboer, 1988; Katz, 1993). In contrast, Cr(III) is nontoxic, essential for human health, and found predominantly in insoluble forms in soils, such as sparingly soluble Cr_2O_3 and $Cr(OH)_3$ (Anderson, 1989; Fendorf, 1995). In 1979, Bartlett and James discovered that certain forms of Cr(III) may oxidize to Cr(VI) in soils, and that Cr(VI) may be reduced to Cr(III) in the same soils. As a result, the opposing solubility and toxicity characteristics of Cr(III) and Cr(VI) and the potential for Cr(III) oxidation in soils represent a unique regulatory challenge for the establishment of protective, health-based clean-up standards for Cr-contaminated soils. Remediation of Cr(VI)-containing soils through reduction to Cr(III) will lower the health and ecological hazards of such soils. Conversely, potential reoxidation of newly formed Cr(III) could increase the hazard of such soils, and thus it must be minimized through remediation processes (James, 1996). Altering the valence state of Cr as a clean-up strategy represents a novel approach to remediation and one that can be successful if based on a knowledge of the forms of Cr involved and the soil properties pertinent to oxidation-reduction reactions.

The early work on Cr(III) oxidation in soils showed that a variable fraction (<15%) of soluble and freshly precipitated Cr(III), such as $CrCl_3$, $Cr(OH)_3$, and some forms of Cr(III) in tannery wastes and sewage sludges, were oxidized to Cr(VI) by soil-borne Mn (III,IV) (hydr)oxides (Bartlett and James, 1979; Amacher and Baker, 1982; James and Bartlett, 1983). Aged precipitates of $Cr(OH)_3$ were less prone to oxidation than were the soluble and freshly precipitated forms, and Cr_2O_3 did not oxidize in laboratory tests (Amacher and Baker, 1982). In addition, reduction reactions of Cr(VI) by organic matter or other reducing agents may occur simultaneously with oxidation of Cr(III) (James and Bartlett, 1983), so the relative rates of oxidation and reduction establish an equilibrium level of Cr(VI) in a particular soil amended with Cr.

Such a balance can be visualized as a seesaw with Mn(III,IV) (hydr)oxides and organic matter on opposite ends (Figure 1). The relative sizes of the circles represent the quantities of the oxidants and reductants for Cr in a particular soil, and their positions with respect to the fulcrum signify their reactivity. The soil pH is a master variable in both reactions (a rolling weight on the seesaw), with maximum oxidation occurring at approximately pH 6 to 7. Reduction of Cr(VI) by organic matter and other electron donors (e.g., Fe(II) and sulfides) is favored by lower pH. Conversely, both oxidation and reduction are inhibited under more alkaline conditions. The arrow pointing down from the fulcrum indicates the level of Cr(VI) found in a particular soil under given redox conditions and at a given pH (James, 1996). The undulation of the seesaw represents the dynamic and adjustable qualities of oxidation-reduction processes in soils and their relevance to remediation-by-reduction strategies for clean-up of Cr-contaminated soils.

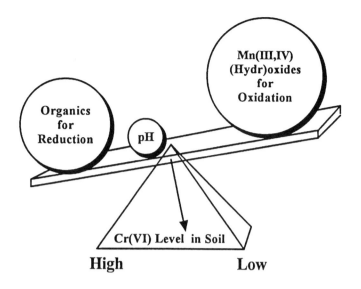

High   **Low**

FIGURE 1

Chromium oxidation-reduction seesaw, the position of which is determined by the relative quantities and reactivity of manganese oxides vs. reducing agents (e.g., organic matter). The soil pH acts as a controllable master variable that helps to set the position of the seesaw and to determine the quantity of Cr(VI) in the soil.

    The solubility and toxicity characteristics of soil-borne Cr(III) and Cr(VI), coupled with the phenomena of Cr(III) oxidation and Cr(VI) reduction, present a continuing challenge for regulatory agencies charged with establishing protective clean-up limits for Cr in soils. In 1991, the U.S. Environmental Protection Agency (USEPA) refused to delist Cr_2O_3 as a hazardous material for disposal in soils because of its presumed potential for oxidation to Cr(VI), an action designed to protect human health and based on the research of Bartlett and James (1979) (Federal Register, 1991). However, Cr_2O_3 is inert to oxidation in soils or under the conditions of the alkaline extraction method for total Cr(VI) in soils (Amacher and Baker, 1982; Vitale *et al.*, 1994). The possibility and extent of oxidation of Cr(III) to Cr(VI) depend on the level and mineralogy of Mn (III,IV) (hydr)oxides, on soil pH; and, most importantly, on the form and solubility of Cr(III) (Bartlett and James, 1979; Fendorf and Zasoski, 1991; James and Bartlett, 1983; Milačič and Štupar, 1995). Under field conditions, Cr(III) oxidation is less likely to occur than under ideal conditions of the laboratory because aged waste materials containing Cr(III) will be less soluble and more inert toward oxidation, especially as $Cr(OH)_3$ precipitates may form on Mn(III,IV)(hydr)oxide surfaces (James and Bartlett, 1983; Fendorf *et al.*, 1992; Fendorf, 1995). In addition, levels of organic C-based materials, Fe(II), sulfides, and other reducing agents in soils create conditions favorable for reduction of Cr(VI) to Cr(III) (Masscheleyn *et al.*, 1992; Vitale *et al.*, 1997).

In its proposed Hazardous Waste Identification Rule (Federal Register, 1995b), USEPA indicated its intent to regulate Cr(III)-containing wastes as if all forms of Cr are as mobile as Cr(VI) and as toxic as soluble Cr(VI). Risk assessments and sorption models in this proposed rule were developed using Cr(VI) or Cr(III) separately, but proposed levels of Cr above which a waste is classified as "hazardous" would be based on total Cr. This approach is based on the assumption that all forms of Cr(III) could potentially oxidize to soluble Cr(VI) in soils. Such an approach to regulating Cr-bearing wastes in soils could mean that many wastes now excluded from regulation under the Resource Conservation and Recovery Act (RCRA) [40 CFR 261.4 (b)(6)] would be considered hazardous.

In a final rule issued the same year (Federal Register, 1995a), USEPA deleted the current land application pollutant limits for Cr in sewage sludge (biosolids) because Cr exists predominantly as Cr(III) in such wastes, and the potential for uptake by plants and oxidation to Cr(VI) were presumed to be low. The cumulative land application limit had been 3000 kg Cr/ha, based on data showing no effects on plant growth at this maximum level of Cr used in a field study on plant growth. These two regulatory actions indicate a dichotomous approach to regulations concerning soil chromium chemistry within different programs of USEPA.

There are two types of hazardous waste (Federal Register, 1995b). "Characteristic" hazardous wastes include those that do not pass the Toxic Characteristic Leaching Procedure, from which the leachate contains >5 mg total Cr/l. However, these wastes may be treated to meet this criterion, thereby rendering them nonhazardous. A "listed" hazardous waste contains Cr at its source of generation, and, as defined, it cannot be delisted without implementing a formal delisting process through USEPA. Any soil that is mixed with a listed, Cr-containing hazardous waste becomes and remains a hazardous waste itself, according to the mixture rule (Federal Register, 1995b).

Accordingly, Cr valence state and speciation need to be evaluated before regulating Cr(III)-containing materials that have minimal or no propensity to oxidize to Cr(VI). In its recent preamble to the proposed Hazardous Waste Identification Rule, USEPA stated that

> Speciation and associated solubility of metal species in wastes which contain metals are key factors that influence the bioavailability of metals. The Agency had no information on the speciation, solubility, or availability of the metals in the wastes in which they are disposed or how they may transform in the environment. The Agency assumed that the metals were in a soluble form, mobile, and available. In the absence of this information, the Agency assumed that the metals are soluble, mobile, and bioavailable. The Agency seeks comment on this approach, and requests data on the speciation and solubility of metals in wastes, together with the conditions of the waste (e.g., pH) that could be disposed by the methods considered in this rulemaking and methodologies that account for the transformation of the metals through changing environmental conditions.

(Federal Register, 1995b, p. 66363)

Therefore, inasmuch as a fraction of certain forms of Cr(III) *can* oxidize in soils, how should this fact be used in setting clean-up standards and in assessing the "hazard" of particular "nonhazardous" Cr(III)-bearing wastes that have been added to certain soils? The position of the seesaw (Figure 1) can be managed in both regulatory and chemical senses to minimize Cr(III) oxidation and maximize Cr(VI) reduction (James, 1996).

This article proposes a new system to quantify the probability of oxidation of different forms of Cr(III) in particular soils and discusses how such a system can be used to establish health-based regulations for Cr in soils. Remediation-by-reduction strategies are also discussed as approaches to diminish the health and ecological hazards of Cr by irreversibly reducing Cr(VI) to immobile forms of Cr(III) without affecting the total Cr concentration. It is essential to recognize that remediation to alter the valence state of a metal is a nontraditional approach to remedial action for metals other than Cr, primarily because Cr is unique in its significant differences in toxicological and chemical behavior between Cr(III) and Cr(VI). For remediation to be successful, potential reoxidation of newly formed Cr(III) must be minimized (James, 1996). Similarly, it is the opposing solubility and toxicity characteristics of Cr(III) and Cr(VI) that make standard setting for Cr-contaminated soils such a challenge for regulators.

OXIDATION AND REGULATION OF CR(III) IN WASTE-SOIL MIXTURES

Predicting the potential formation of Cr(VI) in soils amended with Cr(III) wastes can be based on simple, routine laboratory tests that assign values to four parameters that govern the valence state of Cr. These tests couple (1) a knowledge of the form and solubility of Cr(III) in a waste material, (2) the potential oxidation by the soil of a soluble salt of Cr(III), (3) the potential reduction of soluble Cr(VI) by the soil, and (4) the pH of the soil-waste mixture as a master variable, along with Eh, controlling the balance of oxidation and reduction reactions. The data from these tests can then be used to assign a ranking for each parameter from 0 to 10 (from −10 to 10 for potential Cr(VI) reduction), with lower values assigned to soil or waste conditions that favor lower concentrations of Cr(VI) in a soil (e.g., less Cr(III) oxidation or more Cr(VI) reduction). The sum of the values derived from the four tests is the Potential Chromium Oxidation Score (PCOS) , and it ranges from −10 to 40 for a particular waste-soil combination. Calculated PCOS values from −10 to 10 signify that no upper clean-up limit for Cr(III) disposal in soil is necessary (i.e., Cr(III) is not regulated in that soil), scores from 11 to 20 signify the need for a "high" concentration limit for Cr(III) (e.g., 100 g/kg), and those from 21 to 30 indicate a greater likelihood of oxidation and the prudence of setting a "low" concentration limit for Cr(III) (e.g., 10 g/kg). Scores from 31 to 40 suggest a high probability for Cr(III) oxidation and the persistence of Cr(VI), necessitating remediation or strict engineering controls to minimize potential human contact or exposure.

Because the tests for obtaining a PCOS value are not routinely conducted in commercial environmental laboratories, they are described below. All have been used in research, and they are relatively quick, accurate, and precise procedures.

Waste Oxidation Potential (WOP)

To estimate the potential oxidation of Cr(III) in a waste, a knowledge of the form and solubility of Cr(III), and the Cr(VI) content, in the waste to be evaluated is obtained. Soluble salts of Cr(III), such as $Cr(NO_3)_3$, oxidize more rapidly and extensively than do insoluble compounds, such as $Cr(OH)_3$, and aging and crystallization of the solid phase further inhibit oxidation (James and Bartlett, 1983). Therefore, a score of 10 is assigned to soluble salts of Cr(III) and 0 is assigned to Cr_2O_3 because its oxidation has been shown not to occur in soils, even those high in reactive Mn(III,IV)(hydr)oxides (Amacher and Baker, 1982; James et al., 1995). Assign a value of 3 to wastes with Cr(III) in forms similar to aged $Cr(OH)_3$ (e.g., tannery sludge) and 6 to those similar to freshly precipitated $Cr(OH)_3$ (Figure 2a). Other values may be assigned, relative to these forms of Cr(III). For wastes with unknown forms of Cr(III), a knowledge or identification of the mineral species may be necessary to assign a value for this parameter.

Soil Oxidation Potential for Cr(III) (SOP)

Add 50 ml of 1.0 mM $CrCl_3$ to the field- moist equivalent of 5.0 g air-dried soil in a 250-ml glass beaker, swirl in equilibrium with the atmosphere for 1 h, filter or centrifuge, and measure soluble Cr(VI) in the filtrate (Bartlett and James, 1979). Assign an SOP value equal to 0.5 times the percentage of added Cr(III) oxidized to Cr(VI) (Figure 2b).

Soil Reduction Potential for Cr(VI) (SRP)

Add 50 ml of 0.1 mM K_2CrO_4 to the field- moist equivalent of 5.0 g of air-dried soil in a 250-ml glass beaker, swirl in equilibrium with the atmosphere for 24 h, filter or centrifuge, and measure soluble Cr(VI) in the filtrate. Assign an SRP value based on the equation of Figure 2c.

Soil-Waste pH Modification Value (PMV)

Add 10 ml of 10 mM $CaCl_2$ to the field-moist equivalent of 5.0 g of air-dry soil mixed with 5.0 g of waste in a 25-ml beaker or small plastic cup, stir thoroughly

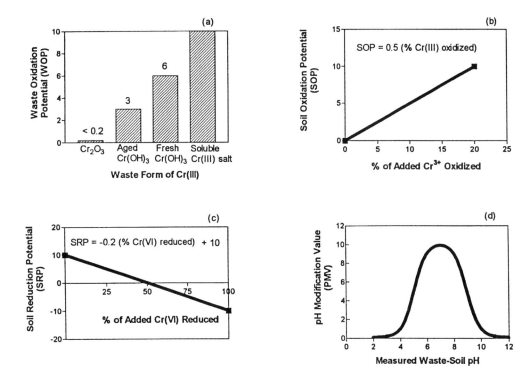

FIGURE 2

Graphical representations of the additive parameters for a Potential Chromium Oxidation Score (PCOS): (a) waste oxidation potential (WOP), (b) soil oxidation potential (SOP), (c) soil reduction potential (SRP), and (d) pH modification value (PMV). PCOS = WOP + SOP + SRP + PMV.

for 1 min, and allow the suspension to stand for 20 min. Submerge a combination pH electrode into the supernatant liquid just until the liquid junction on the reference electrode is covered with liquid. Record pH when the reading is stable. Assign a value of 10 for a pH of 7.0 ± 0.2, a favorable condition for the oxidation of Cr(III) and the persistence of Cr(VI), and assign values between 0 and 10 for pH values greater and less than 7.0 ± 0.2, using the Gaussian distribution of PMV vs. pH (Figure 2d). The hypothetical PMV vs. pH function (Figure 2d) is based on the concept that at pH values < 7, oxidation of Cr(III) is less favorable and Cr(VI) reduction is more favorable than at pH 7. Also, at pH values > 7, both oxidation and reduction are disfavored due to changes in the reactivity of Mn(III,IV) (hydr)oxides and reducing agents.

The sum of WOP, SOP, SRP, and PMV is the Potential Chromium Oxidation Score (PCOS). Examples of calculations of PCOS values for four soils containing four Cr(III)-bearing wastes illustrate a range of values that might be obtained using this method while assessing the potential hazard of a particular form of Cr(III) in a given soil (Table 1). The PCOS approach to regulating Cr in soils also allows flexibility in making changes in a waste or soil property to lower the PCOS value to be able to dispose of a Cr-bearing waste in a particular soil environment. For example, if a PCOS value of 30 were obtained for a Cr(III) soluble salt in a high Mn soil at pH 7, the PCOS value could be decreased by raising or lowering the pH (lowering PMV), by precipitating the Cr(III) and aging the precipitate (lowering WOP), and by lowering the SRP and SOP values through addition of reducing agents for Cr(VI).

The PCOS concept is presented as a flexible investigative and regulatory tool that can be used and discussed as strategies are developed for setting limits for Cr-containing wastes in selected soils. It addresses the important issue of the potential oxidation of Cr(III) in specified waste-soil combinations, and it recognizes and quantifies the balance between oxidation and reduction in predicting levels of Cr(VI) that may be found in Cr-enriched soils.

REDUCTION AND REMEDIATION OF CR(VI) IN SOILS

Remediation-by-reduction strategies for Cr(VI) in soils are based on the concept that Cr(III) has negligible toxicity and minimal mobility compared with Cr(VI), and, therefore, reduction of Cr(VI) to Cr(III) is a means of eliminating the hazard associated with Cr contamination in a soil without changing its total Cr concentration. Successful strategies for effecting such remediation must satisfy three criteria: (1) newly reduced Cr(III) is in a form that is inert to reoxidation, (2) extraneous or objectionable byproducts of the Cr(VI) reduction reaction are not released (e.g., nitrate, Fe(III) oxide crusts, or partially oxidized S-containing compounds), and (3) oxidation-reduction and acidity conditions of the soil are controlled so that Cr(III), other metals, or commingled soil pollutants are not solubilized. In addition, marked changes in these conditions could inhibit the desired Cr(VI) reduction reaction.

Elemental Fe (e.g., steel wool) reduces Cr(VI) to Cr(III) in chromite ore-processing residue-enriched soils and other aqueous environments (James, 1994; Powell *et al.*, 1995), and in the process, alkalinity is generated (Figure 3). The Fe(III) and Cr(III) formed by the redox reaction may create a mixed oxide with low solubility, thereby minimizing the chances for reoxidation of Cr(III). In contrast, reduction of Cr(VI) coupled to oxidation of Fe(II) compounds may generate net acidity, and the soil pH may decrease markedly, depending on the pH buffer capacity of the soil (James, 1994). Similar mixed Cr(III)-Fe(III) oxides may form, however. The reduction of Cr(VI) by organic compounds (e.g., hydroquinone in Figure 3) will also generate alkalinity, and newly formed Cr(III) may be complexed with organic materials, such as humic acids in soils. Such complexation in insoluble forms will inhibit reoxidation if Cr(III) is bound in insoluble organic complexes (e.g., with humic acids). In contrast, solubilization of Cr(III) by organic

TABLE 1
Hypothetical Potential Chromium Oxidation Scores (PCOS) for Four Soils Amended with Four Cr(III)-Bearing Wastes

Waste type	WOP[a]	Quartz sand, pH 8				High Mn soil, pH 7				Alkaline soil, pH 9				Anoxic sediment, pH 7			
		SOP[b]	SRP[c]	PMV[d]	PCOS[e]	SOP	SRP	PMV	PCOS	SOP	SRP	PMV	PCOS	SOP	SRP	PMV	PCOS
Cr_2O_3	0	0	10	8	18	10	0	10	20	2	10	1	13	0	-10	10	0
Aged Cr(OH)$_3$	3	0	10	8	21	10	0	10	23	2	10	1	16	0	-10	10	3
Fresh Cr(OH)$_3$	6	0	10	8	24	10	0	10	26	2	10	1	19	0	-10	10	6
Soluble Cr(III) Salt	10	0	10	8	28	10	0	10	30	2	10	1	23	0	-10	10	10

Note: Higher values indicate a greater probability for oxidation of Cr(III) or persistence of Cr(VI) in a waste-soil combination.

[a] WOP = waste oxidation potential (Figure 2a); applicable to each of the four soils.

[b] SOP = soil oxidation potential (Figure 2b).

[c] SRP = soil reduction potential (Figure 2c). Values < 0 indicate that soil reduction decreases the Cr(VI) level.

[d] PMV = pH modification value (Figure 2d).

[e] PCOS = WOP + SOP + SRP + PMV.

REDUCTION REACTIONS: EXAMPLES

REDUCTION BY <u>ELEMENTAL IRON</u> [with formation of Fe(III)]

$$Fe \; + \; CrO_4^{2-} \; + \; 0.5 \; H_2O \; + \; 2 \, H^+ \; \rightarrow \; Fe(OH)_3 \; + \; 0.5 \; Cr_2O_3$$

REDUCTION BY <u>DIVALENT IRON</u> [with formation of Fe(III)]

$$6 \; Fe^{2+} \; + \; 2 \; CrO_4^{2-} \; + \; 13 \; H_2O \; \rightarrow$$

$$6 \; Fe(OH)_3 \; + \; Cr_2O_3 \; + \; 8 \; H^+$$

REDUCTION BY <u>ORGANIC COMPOUNDS</u>, e.g., HYDROQUINONE [with formation of quinone]

$$1.5 \; C_6H_6O_2 \; + \; CrO_4^{2-} \; + \; 2 \; H^+ \; \rightarrow$$

$$0.5 \; Cr_2O_3 \; + \; 1.5 \; C_6H_4O_2 \; + \; 2.5 \; H_2O$$

FIGURE 3

Reduction reactions for Cr(VI) that illustrate the stoichiometry and alkalinity changes associated with the use of Fe, Fe(II), and organic compounds for remediation.

acids may increase oxidation of Cr(III), depending on the form of Cr, pH, and the organic acid involved (James and Bartlett, 1983). The most efficacious methods for remediating Cr(VI)-contaminated soils will depend on native soil pH, the presence of natural reducing agents, forms of Cr(VI) present (e.g., their solubility), the pH buffer capacity of the soil, and the ease with which the reducing agents may be incorporated into depths of the soil where Cr(VI) is found.

SOIL CHROMIUM REGULATION AND REMEDIATION IN PERSPECTIVE

Chromium is a constituent of many products valued by society, and it has myriad uses in modern industrial processes. Examples are stainless steel, plated metals, tanned leather, pressured-treated lumber, textile dyes and mordants, pigments, ceramic glazes, refractory bricks, and as the trace element responsible for the brilliant red and green colors of the ruby and emerald (Petrucci and Harwood, 1993). The environmentally sound disposal of Cr wastes from the production of many of these commodities and innovative strategies for cleaning up historic, waste-enriched soils high in Cr(VI) will require attention to the oxidation-reduction chemistry of Cr in soils. This is a concept that is unique to Cr because of the widely different toxicity characteristics and environmental behavior of its two most common valence states.

Assumptions that complete oxidation of all forms of Cr(III) will occur and that Cr(VI) will be completely reduced to Cr(III) if the wastes are mixed with soils are untenable and not based on scientific findings. Regulations to limit exposure to Cr(VI) must consider the implications of designating particular Cr-bearing wastes as hazardous from both toxicological and remediation perspectives. The potential to control the valence state of Cr by chemical transformation should facilitate redefinition and reclassification of "hazardous" wastes as nonhazardous through the beneficial use of redox chemistry and the PCOS approach.

Additionally, the PCOS system for quantifying and predicting potential oxidation of specific forms of Cr(III) in particular soil environments can be used to ascertain the hazard of a particular waste-soil combination. A knowledge of the reduction chemistry of Cr(VI) in soils also is needed to design predictable remediation-by-reduction strategies to "clean" Cr(VI)-contaminated soils. Recognition of the importance of the different forms, chemical reactions, mobility, and bioavailability of Cr(III) and Cr(VI) can be the basis for sound regulatory and remedial actions to protect human health from Cr(VI) in soil-waste environments.

REFERENCES

Amacher, M. C. and Baker, D. E. 1982. Redox reactions involving chromium, plutonium, and manganese in soils. US Dept. of Energy Final Rep. DOE/DP/04515-1. Institute for Research on Land and Water Resources, Pennsylvania State Univ., University Park, PA. pp.1–166.

Anderson, R. A. 1989. Essentiality of chromium in humans. *Sci. Tot. Environ.* **86,** 75–81.

Bartlett, R. and James, B. 1979. Behavior of chromium in soils. III. Oxidation. *J. Environ. Qual.* **8,** 31–35.

Federal Register. 1991. **56,** 58859.

Federal Register. 1995a. **60,** 54764.

Federal Register. 1995b. **60,** 66344.

Fendorf, S. 1995. Surface reactions of chromium in soils and waters. *Geoderma* **67,** 55–71.

Fendorf, S., Fendorf, M., Sparks, D. L., and Gronsky, R. 1992. Inhibitory mechanisms of Cr(III) oxidation by δ-MnO$_2$. *J. Coll. Inter. Sci.* **153,** 37–54.

Fendorf, S. and Zasoski, R. J. 1991. Chromium(III) oxidation by δ-MnO$_2$. I. Characterization. *Environ. Sci. Technol.* **26,** 79–85.

James, B. R. 1994. Hexavalent chromium solubility and reduction in soils enriched with chromite ore processing residue. *J. Environ. Qual.* **23,** 227–233.

James, B. R. 1996. The challenge of remediating chromium-contaminated soil. *Environ. Sci. Technol.* **30**, 248A-251A.

James, B. R. and Bartlett, R. J. 1983. Behavior of chromium in soils. VI. Interactions between oxidation-reduction and organic complexation. *J. Environ. Qual.* **12**, 173–176.

James, B. R., Petura, J. C., Vitale, R. J., and Mussoline, G. R. 1995. Hexavalent chromium extraction from soils: a comparison of five methods. *Environ. Sci. Technol.* **29**, 2377–2381.

Katz, S. A. 1993. The toxicology of chromium with respect to its chemical speciation. *J. Appl. Toxicol.* **13**, 217–224.

Masscheleyn, P. H., Pardue, J. H., DeLaune, R. D., and Patrick, W. H., Jr. 1992. Chromium redox chemistry in a Lower Mississippi Valley bottomland hardwood wetland. *Environ. Sci. Technol.* **26**, 1217–1226.

Milačič, R. and Štupar, J. 1995. Fractionation and oxidation of chromium in tannery waste- and sewage sludge-amended soil. *Environ. Sci. Technol.* **29**, 506–514.

Petrucci, R. H. and Harwood, W. 1993. *General Chemistry.* New York, MacMillan.

Powell, R. M., Puls, R. W., Hightower, S. K., and Sabatini, D. A. 1995. Coupled iron corrosion and chromate reduction: Mechanisms for subsurface remediation. *Environ. Sci. Technol.* **29**, 1913–1922.

Vitale, R. J., Mussoline, G. R., Petura, J. C., and James, B. R. 1994. Hexavalent chromium extraction from soils: Evaluation of an alkaline digestion method. *J. Environ. Qual.* **23**, 1249–1256.

Vitale, R. J., Mussoline, G. R., Rinehimer, K. A., Petura, J. C., and James, B. R. 1997. Extraction of sparingly soluble chromate from soils: evaluation of methods and Eh-pH effects. *Environ. Sci. Technol.* **31**, 390–394.

Yassi, A., and Nieboer, E. 1988. In: *Chromium in Natural and Human Environments.* pp. 443–495. (Nriagu, J. O. and Nieboer, E., Eds.) New York, Wiley-Interscience.

Journal of Soil Contamination, 6(6):581–593 (1997)

Cr(VI) Soil Analytical Method: A Reliable Analytical Method for Extracting and Quantifying Cr(VI) in Soils

Rock J. Vitale,[1] George R. Mussoline,[1] John C. Petura,[2] and Bruce R. James[3]

[1]Environmental Standards, Inc., Valley Forge, Pennsylvania; [2]Applied Environmental Management, Inc., Malvern, Pennsylvania; and [3]University of Maryland, College Park, Maryland

Chromium has been used in the industrialized world in many applications for more than a century. Chromium is a trace metallic element found in the Earth's crust, and when it is found in concentrated ore deposits it is principally as $FeCr_2O_4$ (chromite ore). In the environment, chromium is typically found in the trivalent and hexavalent states. These two oxidation states have differing toxicities and mobilities. Hexavalent chromium [Cr(VI)] is classified as a known human carcinogen (via inhalation) and is rather mobile, whereas trivalent chromium [Cr(III)] is comparatively benign (it is an essential dietary element for humans) and relatively immobile. This significant toxicological and geochemical disparity between the two valence states necessitates that environmental investigators be able to quantitatively distinguish between these two forms in solid environmental media. Several regulatory-approved analytical techniques exist for the quantitative differentiation between Cr(VI) and Cr(III) in aqueous solutions and for the measurement of total Cr in the solid phases. However, until recently, a regulatory-approved analytical technique for the extraction of total Cr(VI) from soils and solid phases did not exist. Previous research was evaluated in order to develop an accurate and precise method for extracting and analyzing total Cr(VI) from solid media without causing Cr(III) oxidation or Cr(VI) reduction during the extraction and analytical process. Incomplete Cr(VI) spike recoveries [viz., reduction of Cr(VI) to Cr(III)] and possible oxidation of Cr(III) to Cr(VI) during the Cr(VI) extraction and analysis process have been suggested as insurmountable issues that precluded the development of a defensible extraction technique. Extensive research and an elaborate method evaluation study using a hot alkaline extractant combined with routine redox characterization of test samples resulted in the development of a reliable extraction method for the analysis of total Cr(VI). Based on that research, in June 1997 the U.S. EPA promulgated SW-846 Method 3060A for inclusion in the Third Update to the Test Methods for Evaluating Solid Wastes, *SW-846, 3rd ed. The proposed SW-846 Method 3060A provides procedural detail for method utilization with significant guidance to environmental investigators on sample redox characterization, method limitations, and the interpretation of matrix spike data.*

1058-8337/97/$.50

*C*HROMIUM (Cr) is a naturally occurring element typically found in U.S. soils from 1 to 2000 mg kg^{-1}, apart from Cr-rich ore deposits that have been shown to contain up to 27% Cr (Copson, 1956; Dragun and Chiasson, 1991). Chromium can exist in multiple valence states, with the 3^+ and 6^+ being the most frequently encountered, and the 2^+ valence being relatively unstable and rarely encountered unless specifically produced (Nriagu and Nieboer, 1988). Trivalent Cr is ubiquitous in the environment and occurs naturally, while almost all known sources of Cr(VI) are derived from human activities (World Health Organization, 1988). Trivalent Cr found in the environment is typically found as insoluble crystalline or para crystalline compounds such as chromium hydroxide ($Cr(OH)_3$) and chromium (III) oxide (Cr_2O_3). Trivalent Cr is relatively unreactive compared with Cr(VI) and is considered an essential nutrient for humans (IRIS, 1993; Anderson, 1989).

Hexavalent chromium has been commercially prepared for industrial uses for many decades by alkaline high-temperature roasting, and subsequent leaching with H_2SO_4 from $FeCr_2O_4$, chromite ore. The single most important use of Cr is for stainless steel and other metal alloys. Chromates and dichromates of sodium and potassium, and chromium alums of potassium and ammonium represent the common commercially available Cr(VI) compounds (Weast, 1978). The various forms and uses of both Cr(III) and Cr(VI) and the significant diversity of uses are provided (Table 1). As with any ore-extraction process used over decades within an industrialized society, slag and various Cr-containing byproducts were released to the environment during a time when the health impacts of such actions were not known or considered.

Since the mid-1900s, when the potential human health impacts of environmental exposure to Cr(VI) were recognized, Federal and State regulatory agencies were charged with minimizing human exposure to a variety of substances, including Cr(VI). Further research on the potential toxicological effects of Cr(VI) resulted in the classification of Cr and its compounds as hazardous wastes under the Resource Conservation and Recovery Act (RCRA) and hazardous substances under the Clean Water Act, the Clean Air Act, and the Comprehensive Environmental Resource Compensation and Liability Act (CERCLA). Hexavalent chromium and chromic acid have been ranked 18th and 203rd, respectively, on the CERCLA Hazardous Substance List based on the frequency of occurrence, toxicity, and potential for human exposure (Hazardous Waste Consultant, 1993).

Based on the lack of a regulatory-approved Cr(VI) extraction method, there has been a tendency, until recently, among some regulatory agencies to base soil remediation criteria on the analysis for non-valence specific total Cr and conservatively assume the results to represent the concentration of Cr(VI) when performing a toxicological assessment of a site. Recognizing the significant disparity between the toxicity of Cr(III) and Cr(VI), there has been broad interest among the

TABLE 1
Uses of Selected Chromium Compounds[a]

Chromium compound	Chemical formula	Cr valence	Description	Water solubility	Uses
Chromic acid	CrO_3	6+	Dark purplish – red crystals	S	Chemicals (chromates, oxidizing agents, catalyst), chromium plating intermediates, anodizing, ceramic glazes, textile mordant, etchant for plastics, metal cleaning, inks, dyes, paints, colored glass
Chromic chloride	$CrCl_3$	3+	Violet crystals	I	Chromium plating, oletins polymerization catalyst, waterproofing, textile mordant.
Chromic hydroxide	$CrCl_3 \cdot 6H_2O$	3+	Greenish-black or violet deliquescent crystals	S	Guignet's green, catalyst, tanning agent, textile mordant
	$Cr(OH)_3$	3+	Green gelatinous precipitate	I	
Chromic oxide	Cr_2O_3	3+	Bright green extremely hard crystals	I	Metallurgy, green paint pigment, ceramics, organic synthesis catalyst, green granules in asphalt roofing, abrasive, refractory, brick component
Chromic sulfate	$Cr_2(SO_4)_3$	3+	Violet or red powder	I	Chromium plating, chromium alloys, catalyst, green paint, varnishes and ink, ceramic glazes, leather tanning, textile mordant
	$Cr_2(SO_4)_3 \cdot 15\ H_2O$	3+	Dark green amorphous scales	S	
	$Cr_2(SO_4)_3 \cdot 18H_2O$	3+	Violet cubes	S	
Chromium	Cr	0	"Hard, brittle semi-gray metal"	I	Alloying and plating element on metal and plastic, substrate for corrosion resistance, nuclear and high-temperature research, inorganic pigment constituent
Chromium ammonium sulfate	$CrNH_4(SO_4)_2 \cdot 12H_2O$	3+	Green powder or deep violet crystals	S	Leather tanning, textile mordant
Chromous chloride	$CrCl_2$	2+	White deliquescent needles	S	Reducing agent, catalyst, reagent
Chromous sulfate	$CrSO_4 \cdot 5H_2O$	2+	Blue crystals	S	Oxygen scavenger, reducing agent, analytical reagent.
Lead chromate	$PbCrO_4$	6+	Yellow crystals	I	Pigments in industrial paints, rubber, plastics, ceramic coatings, organic analysis
Potassium chromate	K_2CrO_4	6+	Yellow crystals	S	Analytical reagent, aniline black, textile mordant, enamels, pigment in paints, inks

TABLE 1 (continued)
Uses of Selected Chromium Compounds[a]

Chromium compound	Chemical formula	Cr valence	Description	Water solubility	Uses
Potassium dichromate	$K_2Cr_2O_7$	6+	Bright yellowish red transparent crystals	S	Oxidizing agent, analytical reagent, brass pickling compositions, electroplating, pyrotechnics, explosives, safety matches, textiles, dyes and inks, glues and adhesives, leather tanning, wood stains, paint pigments, synthetic perfumes, ceramics, bleaching fats and waxes
Sodium chromate	$Na_2CrO_4 \cdot 10H_2O$	6+	Yellow translucent efflorescent crystals	S	Inks, dyes, paint pigment, leather tanning, iron corrosion protection, wood preservatives
Sodium dichromate	$Na_2Cr_2O_7 \cdot 2H_2O$	6+	Red or orange deliquescent crystals	S	Colorimetry (copper determination), complexing agent, oxidation inhibitor in ethyl ether

Note: I - Insoluble.

[a] Source: Hawley's Condensed Chemical Dictionary, 12th edition, 1993.

regulated community in accurately and precisely differentiating Cr(VI) quantitatively from Cr(III). Successful valence differentiation and quantitation are necessary in order to perform realistic assessments of risk to potential receptors exposed to Cr and to remediate sites to Cr(VI) concentrations that are below applicable health-based action limits.

ANALYTICAL METHODOLOGY

There are several USEPA-approved methods that differentiate between Cr(III) and Cr(VI) in solution. Total Cr analysis is typically performed by graphite furnace atomic absorption (GFAA) or inductively coupled plasma (ICP) atomic emission spectroscopy (USEPA, 1983; USEPA, 1996). For the determination of Cr(VI) in airborne particulates, a method was developed (ASTM D5281-92) for collecting airborne particulate matter in an alkaline impinger solution with the analysis by ion chromatography/visible absorption spectroscopy via a post-column colorimetric reaction with diphenylcarbazide (ASTM, 1992). Diphenylcarbazide has been used as a selective color reagent for the detection of Cr(VI) in solution for a number of years (Bartlett and James, 1979; Cazeneuve, 1900). There are several regulatory approved analytical methods for the analysis of Cr(VI) in aqueous samples (i.e., SW-846 Method 7195–Cr(VI) Coprecipitation; SW-846 Method 7196A–Cr(VI) Colorimetric; SW-846 Method 7197–Cr(VI) Chelation Extraction; SW-846 Method 7198–Cr(VI) Differential Pulse Polarography; and SW-846 Method 7199–Cr(VI) Ion Chromatography). Method 7196A and Method 7199 represent direct measurements of Cr(VI), while Method 7195, Method 7197, and Method 7198 are used less commonly as they represent indirect measurement techniques for Cr(VI).

An alkaline extraction procedure, SW-846 Method 3060, for the preparation of soil samples for analysis for total Cr(VI) by one of the above analytical methods, was used for a number of years until 1986. In 1986, a USEPA-funded research study did not achieve consistent results among samples with SW-846 Method 3060 (USEPA, 1984; USEPA, 1986). The researchers concluded that the Cr oxidation state is matrix specific and may be unstable and unpredictable (in environmental samples) once it is solubilized in either an acidic or basic aqueous extraction medium. As a result the method was deleted from the third edition of the USEPA *Test Methods for Evaluating Solid Wastes*. A replacement method for the extraction of total Cr(VI) in solid media (Vitale *et al.*, 1993) was promulgated by the USEPA in 1997 (USEPA, 1997) .

The perceived problems associated with the 1986 USEPA-funded research of SW-846 Method 3060 stemmed from a unique characteristic that is not encountered when analyzing samples for total metals. Because Cr(VI) is a strong oxidant, there are certain conditions in the environment that preclude Cr from existing in the 6^+ valence state. Under highly reducing conditions (e.g., anoxic, sulfidic sediments), Cr(VI) matrix spikes are rapidly reduced to insoluble Cr(III) (James and Bartlett, 1983b; Massacheleyn *et al.*, 1992; Vitale *et al.*, 1994; Vitale *et al.*,

1995). Unfortunately, zero or low (<75%) percent matrix spike recoveries for Cr(VI), previously observed in such reducing samples, were interpreted in the same manner as were total metals and determined to represent "method failure." A more appropriate interpretation of these low matrix spike recoveries for Cr(VI) is that the reduction of the Cr(VI) to Cr(III) should have been investigated and should have been recognized as a predictable soil chemical reaction for specific soil types (James and Bartlett, 1983b).

With the removal of SW-846 Method 3060 in 1986, and faced with the need to speciate between the Cr(III) and Cr(VI) valence states, environmental investigators have previously attempted to utilize various leaching procedures to quantitatively extract Cr(VI) from solid media. These include the USEPA EP Toxicity leaching procedure, the more recent USEPA Toxicity Characteristic Leaching Procedure (TCLP), various deionized water neutral extraction procedures (e.g., ASTM Methods D 3987–85 and D 4646–87), and buffered extractant procedures (phosphate buffer). The results of the latter two extractions have been used to quantify and define operationally soluble and exchangeable forms of Cr(VI) (James and Bartlett, 1983b; James *et al.*, 1995). Recent research has demonstrated that these two leachate methods do not efficiently extract both soluble and insoluble forms of Cr(VI) (James *et al.*, 1995).

Due to a compelling need to extract and quantify total Cr(VI) in solid media, research was conducted over the last several years to modify SW-846 Method 3060 (Vitale *et al.*, 1993). Although the basic chemistry has remained the same, the modifications to SW-846 Method 3060 have enhanced the efficiency of the extraction process principally by reducing the soil sample weight and decreasing the sample weight to volume ratio. SW-846 Method 7196A has also been refined in this process. The conditions of SW-846 Method 3060, USEPA SW-846 Method 3060A, and the analytical portions of SW-846 Method 7196A and Modified Method 7196A are compared (Tables 2 and 3) (USEPA, 1996).

TABLE 2
Differences between
Method 3060 and Method 3060A

Item	Method 3060[a]	Method 3060A
Sample weight (wet)	100 g	2.5 g
Alkaline digest solution volume	400 ml	50 ml
Final digest volume	1000 ml	100 ml
Digestion temperature	Near boiling	90–95°C
Digestion time	30–45 min	60 min
Nitric acid acidification	pH between 7–8	pH between 7–8

[a] As appeared in second edition of SW-846.

586

TABLE 3
Differences between Method 7196 and Method 7196A

Item	Method 7196	Method 7196A
Volume used for analysis	95 ml	45 ml
Amount of diphenylcarbazide (DPC) added	2.0 ml	1.0 ml
Acidification with H_2SO_4	pH to 1.5 to 2.5	pH 1.6 to 2.2
Turbidity	Subtract absorbance observed before DPC addition from absorbance after DPC addition	Same as 7196A, plus filter through 0.45 µm or 0.1 µm membrane
Initial calibration	0.5 to 5.0 mg/l	0.05 to 2.0 mg/l
Continuing calibration	After every 15 samples	After every 10 samples
Blanks	One per batch	One preparation blank per batch; reagent blank after every continuing calibration
Spike duplicate	Every 10 samples (post-digestion) with 85–115% acceptance criterion	A duplicate every 20 samples; a pre-digestion spike (75–125%) and a post-digestion spike (85–115%) per 20 samples
Lab control sample (LCS)	None mentioned	One LCS per 20 samples with acceptance criterion 80–120%

SW-846 Method 3060A involves the use of a hot alkaline (pH 12) solution containing 0.28 M Na_2CO_3 and 0.5 M NaOH, which effectively solubilizes both water-soluble and water-insoluble Cr(VI) (James, 1994; Vitale *et al.*, 1994). Once total Cr(VI) is solubilized, the Cr(VI) is colorimetrically analyzed by adding diphenylcarbazide and adjusting the solution to pH 2 using H_2SO_4. Careful monitoring of the pH prior to sample absorbance readings is one of the refinements that have been made to SW-846 Method 7196A. The analysis of Cr(VI) is either performed manually using a spectrophotometer at 540 nm (SW-846 Method 7196A) or by using ion chromatography (IC) with a post-column derivitization reaction with diphenylcarbazide (SW-846 Method 7199) (USEPA, 1996). For samples that yield turbid or highly colored alkaline extracts, greater sensitivity and lower detection limits can typically be achieved using the IC analysis compared with the manual colorimetric method. The high frequency of acceptable matrix spike recoveries attained using even the sparingly soluble chromate compounds $BaCrO_4$ and $PbCrO_4$ has demonstrated the reliability and robustness of SW-846 Method 3060A (Vitale *et al.*, 1993; Vitale *et al.*, 1994; James *et al.*, 1995). Several thousand non-reducing soil samples have been analyzed to date, with acceptable matrix spike recoveries (75 to 125%).

The collective research that established the basis for SW-846 Method 3060A demonstrated that method-induced reduction of Cr(VI) to Cr(III) did not contribute to low or 0% matrix spike recoveries. Soils that exhibited highly reducing properties could not maintain the Cr(VI) spikes in the 6^+ valence state. SW-846 Method 3060A contains a detailed decision tree (Figure 1) to assist the user in the interpretation of the quality control (QC) data that are needed to substantiate the quantification of the Cr(VI) results. Such QC data are distinctly different from the conventional QC data requirements for total metals analysis, which is also a major enhancement over prior extraction procedures for Cr(VI).

In situations where low or zero percent matrix spike recoveries are observed and a reducing sample is suspected, SW-846 Method 3060A stipulates the measurement of a number of redox-indicating ancillary parameters to assist data users in interpreting Cr(VI) matrix spike recoveries. Four major redox-indicating ancillary parameters include pH (using USEPA SW-846 Method 160.3M), oxidation-reduction potential (ORP) (using ASTM D1498–76), sulfides (using USEPA SW-846 Method 9030), and total organic carbon (TOC) (using USEPA Region II method in which organic compounds are decomposed by pyrolysis in the presence of O_2 or air) (USEPA, 1983; ASTM, 1976; USEPA, 1995; USEPA Region II, 1993, personal communication). The effects of the redox potential on the speciation of Cr in samples needs to be understood or predicted through the use of these ancillary parameters. The Cr(VI) matrix spike recovery data, in conjunction with ancillary data on the redox status of a subject soil sample, provides the user essential information needed to understand and evaluate the properties of the sample with respect to the solubility and redox status of Cr(VI) under field conditions and during chemical analyses (Vitale *et al.*, 1994).

Based on the research performed on a wide variety of samples using SW-846 Method 3060A, and departing from the conventional interpretative approach for QC data for total metals, data associated with low or 0% Cr(VI) matrix spike recoveries should not be dismissed as unreliable, but must be evaluated in accordance with established redox chemistry of Cr in soils or sediments (Adriano, 1986; Bartlett and James, 1988; Rai *et al.*, 1989; Richard and Bourg, 1991). With pH and ORP having such significance with respect to the redox status of a soil or sediment sample, method users are referred to an Eh-pH diagram for $HCrO_4^-/Cr(OH)_3$ (Figure 2), which can be used to assess the redox characteristics of a sample.

With respect to a limitation of SW-846 Method 3060A, method-induced oxidation of Cr(III) to Cr(VI) has been observed in samples demonstrated to contain soluble forms of Cr(III) and exhibit high levels of MnO_2. When subjected to the aerated alkaline conditions of SW-846 Method 3060A, soluble forms of Cr(III) can form a fresh $Cr(OH)_3$ precipitate and $Cr(OH)_4^-$ at pH 12 to 13. This fresh precipitate is then available to be partially oxidized to Cr(VI) under the aerated alkaline conditions of SW-846 Method 3060A. Therefore, such conditions can impart a high bias on the resultant Cr(VI) concentrations observed in soluble Cr(III)-

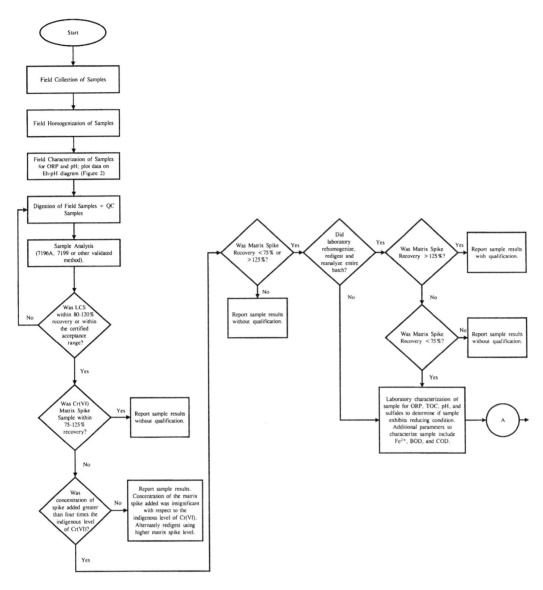

FIGURE 1

Quality control flow chart.

amended soils. However, with the exception of a fresh spill of soluble Cr(III), the soil-borne forms of Cr(III) found in environmental samples are aged, crystalline, $Cr(OH)_3$ and Cr_2O_3, both of which have not been observed to oxidize under the aerated alkaline conditions of SW-846 Method 3060A (James and Bartlett, 1983a; Amacher and Baker, 1982; Zatka, 1985).

The presence of soluble Cr(III) in samples can be approximated by performing a water extraction (ASTM D 4646-87, D 5233-92, or D 3987-85) and analyzing the

589

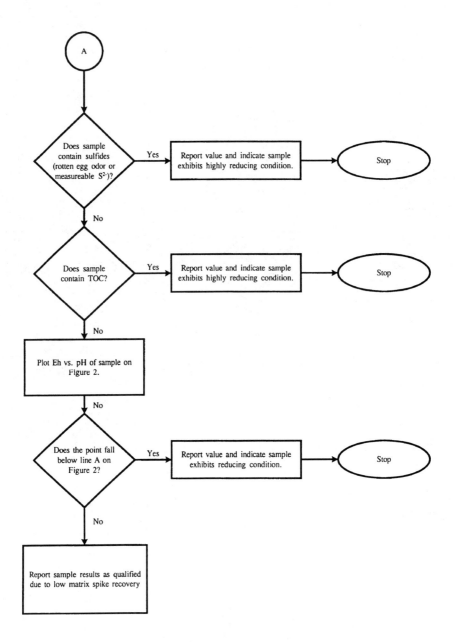

FIGURE 1 (continued)

resultant leachate for both Cr(VI) and total Cr. The difference between the two values approximates the amount of soluble Cr(III) present in a given sample.

When soluble Cr(III) or freshly precipitated $Cr(OH)_3$ is suspected of being present in a sample, SW-846 Method 3060A specifies the addition of Mg^{2+}, which previously has been shown to reduce or eliminate the occurrence of oxidation of Cr(III) to Cr(VI) (Zatka, 1985). This observation was confirmed during the devel-

The dashed lines define Eh-pH boundaries commonly encountered in soils and sediments.

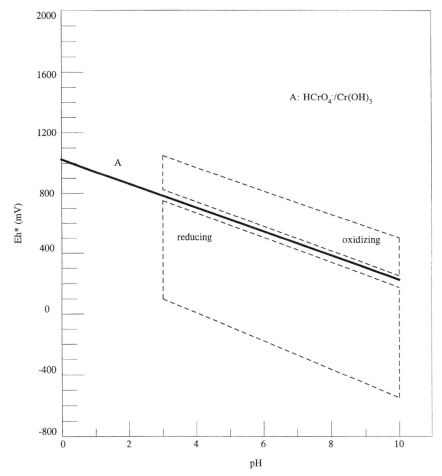

* Note the Eh values plotted on this diagram are corrected for the reference electrode voltage: 244 mV units must be added to the measured value when a separate calomel electrode is used, or 199 mV units must be added if a combination platinum electrode is used.

FIGURE 2

Eh/pH phase diagram.

opment of SW-846 Method 3060A, although there may be additional operative factors in various soil types that remain to be addressed (Vitale *et al.*, 1994).

CONCLUSIONS

SW-846 Method 3060A represents a reliable method for the extraction of total Cr(VI) from solid matrices. Satisfactory matrix spike recoveries (viz., 75 to 125%) have been observed in thousands of non-reducing environmental soils samples.

Evaluation of the proposed SW-846 Method 3060A demonstrated that method-induced reduction is not the cause of low or 0% matrix spike recoveries when they occur in such non-reducing samples. Soils that exhibit highly reducing characteristics (e.g., anoxic sediments) are not capable of maintaining Cr in the $^{+6}$ valence state in an environmental setting or after adding a Cr(VI) matrix spike in the laboratory. In instances where low or 0% matrix spike recoveries are observed, the results of redox-indicating ancillary parameters must be evaluated to assess if the sample in question has the capacity to support Cr(VI). Data obtained using proposed SW-846 Method 3060A on samples that have low or 0% matrix spike recoveries and are demonstrated to be highly reducing should be considered acceptable for use as reported.

Evaluation of the proposed SW-846 Method 3060A has shown that partial method-induced oxidation of Cr(III) to Cr(VI) occurs only when soluble Cr(III) or freshly precipitated $Cr(OH)_3$ is present, which is not likely in environmental samples except those associated with a fresh spill. In the event soluble Cr(III) is suspected or observed, the proposed SW-846 Method 3060A specifies the addition of Mg^{+2}, which has been shown to suppress potential oxidation of Cr(III).

REFERENCES

Adriano, D. C. 1986. *Trace Elements in the Terrestrial Environment*, New York, Springer-Verlag.

Amacher, M. C. and Baker, D. E. 1982. Redox Reactions Involving Chromium, Plutonium, and Manganese in Soils. DOE/DP/04515–1. Inst. for Research on Land and Water Resources. Penn State Univ. and U.S. Dept. of Energy. Las Vegas, NV.

Anderson, R. A. 1989. Essentiality of chromium in humans. *Sci. Tot. Environ.* **86,** 75–81.

ASTM (American Society of Testing and Materials). 1976. Standard Practice for Oxidation-Reduction Potential in Water. *ASTM Designation D1498–76.* Philadelphia, PA.

ASTM (American Society of Testing and Materials). 1992. Standard Test Method for Collection and Analysis of Hexavalent Chromium in Ambient Workplace or Indoor Atmospheres. *ASTM Designation D5281–92.* Philadelphia, PA.

Bartlett, R. and James, B. R. 1979. Behavior of chromium in soils. III. Oxidation. *J. Environ. Qual.* **8,** 31–35.

Bartlett, R. and James, B. R. 1988. Mobility and bioavailabilty of chromium in soils. pp. 267–304. In: *Chromium in the Natural and Human Environments.* (Nriagu, J. O. and Nieboer, E., Eds.) New York, Wiley-Interscience.

Cazeneuve, P. 1900. Sur la diphenylcarbazide, reactif tressensible de quelque composes metalliques: cuivre, mercure, fer au maximum, acide chromique. *Bull. Soc. Chim. Paris* **23,** 701–706.

Copson, R. L. 1956. Production of chromium chemicals. *Chromium: Chemistry of Chromium and Its Compounds,* Vol. I, (Udy, M. J., Ed.) pp. 262–280, New York, Reinhold Publishing.

Dragun, J. and Chiasson, A. 1991. *Elements in North American Soils.* pp. 56–62. Greenbelt, Maryland, Hazardous Materials Control Institute.

Federal Register (1997) **62,** 32452.

Hazardous Waste Consultant. 1993. *EPA Updates CERCLA Priority List of Hazardous Substances.* **10(5).** pp. 2.26–2.30. Lakewood, Colorado, McCoy and Associates, Inc.

IRIS (Integrated Risk Management System). 1993. A continuously updated electronic database maintained by the U.S. Environmental Protection Agency. Bethesda, Maryland, National Library of Medicine.

James, B. R. and Bartlett, R. J. 1983a. Behavior of chromium in soils. VI. Interactions between oxidation-reduction and organic complexation. *J. Environ. Qual.* **12,** 173–176.

James, B. R. and Bartlett, R. J. 1983b. Behavior of chromium in soils. VII. Adsorption and reduction of hexavalent forms. *J. Environ. Qual.* **12,** 177–181.

James, B. R. 1994. Hexavalent chromium solubility and reduction in soils enriched with chromite ore processing residue. *J. Environ. Qual.* **23,** 227–233.

James, B. R., Petura, J. C., Vitale, R. J., and Mussoline, G. R. 1995. Hexavalent chromium extraction from soils: a comparison of five methods. *Environ. Sci. Technol.* **29,** 2377–2381.

Massacheleyn, P. H., Pardue, J. H., DeLaune, R. D., and Patrick, W. H. 1992. Chromium redox chemistry in a lower Mississippi Valley bottomland hardwood wetland. *Environ. Sci. Technol.* **26,** 1217–1226.

Nriagu, J. and Nieboer, E. 1988. *Chromium in the Natural and Human Environments.* pp. 215–229. New York, John Wiley & Sons.

Rai, D., Eary, L. E., and Zachara, J. M. 1989. Environmental chemistry of chromium. *Sci. Total. Environ.* **86,** 15–23.

Richard, F. C. and Bourg, A. C. M. 1991. Aqueous geochemistry of chromium: a review. *Water Res.* **23,** 807–816.

Sax, N. I. and Lewis, R. J., Sr., Eds. 1993. *Hawley's Condensed Chemical Dictionary,* 13th ed. New York, Van Nostrand Reinhold.

U.S. Environmental Protection Agency. 1983. Methods for Chemical Analysis of Water and Wastes. U.S. EPA Rep. 600/4-79-020 U.S. EPA Enviro. Monit. and Support Lab., Office of Research and Development, Cincinnati, Ohio.

U.S. Environmental Protection Agency. 1984. Test Methods for Evaluating Solid Wastes, Physical/Chemical Methods. SW-846 2nd ed. Office of Solid Waste and Emergency Response. Washington, D.C.

U.S. Environmental Protection Agency. 1986. Determination of Stable Valence States of Chromium in Aqueous and Solid Matrices-Experimental Verification of Chemical Behavior. U.S. EPA Rep. 600/4-86/039. U.S. EPA Cincinnati, Ohio.

U.S. Environmental Protection Agency. 1996. Test Methods for Evaluating Solid Wastes, Physica/Chemical Methods. SW-846 3rd ed. Office of Solid Waste and Emergency Response. Washington, D.C.

U.S. Environmental Protection Agency. 1997. Federal Register, Part VI. Rules and Regulations. **62,** 32452–32463.

Vitale, R. J., G. R. Mussoline, J. C. Petura, and B. R. James. 1993. A Method Evaluation Study of an Alkaline Digestion (Modified Method 3060) Followed by Colorimetric Determination (Method 7196A) for the Analysis for Hexavalent Chromium in Solid Matrices. Environmental Standards, Inc., Valley Forge, PA.

Vitale, R. J., Mussoline, G. R., Petura, J. C., and James, B. R. 1994. Hexavalent chromium extraction from soils: evaluation of an alkaline digestion method. *J. Environ. Qual.* **23,** 1249–1256.

Vitale, R. J., Mussoline, G. R., Petura, J. C., and James, B. R. 1995. Hexavalent chromium quantification in soils: an effective and reliable procedure. *Am. Environ. Lab.* **7,** 1, 8–10.

Weast, R. C., Ed., 1978. *CRC Handbook of Chemistry and Physics.* 59th ed. pp. B17-B18. Boca Raton, CRC Press.

World Health Organization. 1988. *Environmental Health Criteria 61: Chromium.* p. 12. Geneva, Switzerland.

Zatka, V. J. 1985. Speciation of hexavalent chromium in welding fumes: interference by air oxidation of chromium. *Am. Ind. Hyg. Assoc. J.* **46,** 327–331.

Journal of Soil Contamination, 6(6):595–648 (1997)

Health-Based Soil Action Levels for Trivalent and Hexavalent Chromium: A Comparison with State and Federal Standards

D. M. Proctor, E. C. Shay, and P. K. Scott

ChemRisk Division, McLaren/Hart, Inc., Two Northshore Center, Suite 100, Pittsburgh, Pennsylvania 15212

As part of the Brownfields initiatives being enacted at both the state and federal levels, environmental regulatory agencies are developing health-based screening or action levels to facilitate the reclamation of unused industrial properties. By the end of 1997, approximately 90% of the states will have either adopted federal values or developed their own non-site-specific action levels. These standards can be applied as default cleanup levels, or alternative remediation standards may be developed based on a site-specific risk assessment. A state and federal survey of cleanup levels for hexavalent and trivalent chromium [Cr(VI) and Cr(III)] indicated a general concurrence of approaches (i.e., most states are using the USEPA standard risk assessment model with upper-bound estimates of exposure and USEPA toxicity criteria), although the proposed values vary by as much as 5 orders of magnitude. To understand the variability and uncertainty in these levels, the USEPA Soil Screening Level (SSL) (1996a) equations were calculated for Cr(III) and Cr(VI) by Monte Carlo analysis to develop probability density functions of health-based action levels (HBALs) for residential and industrial land uses. The lowest HBALs were developed for Cr(VI) for the inhalation of particulates pathway (residential = 892 mg/kg; non-residential = 105 mg/kg); therefore, states and regions that do not consider this pathway may have cleanup standards for Cr(VI) that are not adequately protective of public health. It was determined that Cr(III) HBALs are not necessary (lowest value calculated was 178,000 mg/kg for a residential site) due to the very low toxicity of Cr(III). HBALs for the protection of groundwater are extremely variable, and a tiered approach similar to that developed by the USEPA for the SSL framework, which allows for incorporation of some site-specific information, is most appropriate.

1058-8337/97/$.50

*B*ROWNFIELDS initiatives, which have been enacted by most states to reclaim abandoned industrial properties, have prompted the rapid development of health-based soil screening or "action levels"* to expedite the environmental investigation and cleanup process. Brownfields are abandoned, idle, or underutilized industrial and commercial facilities where expansion or redevelopment is complicated by real or perceived environmental contamination. The USEPA has initiated a Brownfields Action Agenda with the goal of reversing the problems of unaddressed contamination, declining property values, and increased unemployment often found in historical industrial areas (USEPA, 1996b). A 2-year pilot of the USEPA program conducted in Cuyahoga County (Cleveland, Ohio) in 1993–94 has generated dramatic benefits from a $200,000 initial investment by the agency. As a result, $1.6 million in private cleanup funds were leveraged, and $110,000 in private foundation funds were invested in Brownfields redevelopment, creating more than $650,000 in new tax revenue and nearly 100 new jobs (USEPA, 1996a). Seventeen additional pilots are underway currently in cities across the U.S. Part of the USEPA Brownfields Action Agenda has been the development of Soil Screening Levels (SSLs) (USEPA, 1996a). The non-site-specific action levels are designed to streamline the regulatory process so that environmental concerns may be more readily and cost-effectively managed. Cost savings should be realized for both the regulators and the regulated community because much of the regulatory decision making for any particular site can be standardized by a simple comparison of chemical-specific criteria to measured levels of chemicals in soil. Additionally, the burden associated with managing environmental contamination for the private sector should be lessened allowing for a more rapid turnover or reuse of otherwise unused land and facilities. Finally, use of these action levels should increase consistency in risk management decision making within and among each regulatory agency.

Action levels are designed as screening tools or as cleanup goals to be used *in lieu* of a site-specific risk assessment. Although developed for Brownfields redevelopment projects, these action levels are also being applied to other environmental programs, including RCRA and CERCLA. In general, action levels are used: (1) as cleanup standards if no risk assessment is prepared, (2) to screen out chemicals that are not a health concern from site-specific risk assessments, or (3) to justify a no action alternative at sites that do not pose a health concern.

To evaluate the consistency of the action levels and the approaches being implemented, a survey of hexavalent and trivalent chromium [Cr(VI) and Cr(III)]

---

*   These concentrations are most commonly referred to as cleanup levels; however, the terminology from state to state varies widely. Common terms include cleanup standards, cleanup levels, screening levels, and screening criteria. As they all essentially serve as default cleanup levels, or "action levels," for simplification purposes, they are collectively referred to as such throughout the article.

action levels, by state and USEPA region, was performed. The health-based action levels set by state and federal agencies generally utilize the standard USEPA risk assessment model, default or "reasonable maximum exposure" (RME) assumptions, and the USEPA toxicity criteria (e.g., reference doses and cancer slope factors) such that the standards are protective of most land uses and site conditions.

To better assess the range and distribution of appropriate action levels developed with this model, the state and federal action levels were compared with a probability density function (PDF) of health-based action levels (HBALs) developed herein by Monte Carlo (e.g., probabilistic) analysis using the USEPA SSL risk assessment equations and published estimates for the range and frequency of input exposure parameters (USEPA, 1996a; Finley et al., 1994). Monte Carlo or probabilistic analyses are frequently used in health risk assessments to provide more information on the variability and uncertainty of the risk estimates (Finley et al., 1994). An advantage of using probabilistic methods is that the effects of compounded conservatism in exposure parameter point estimates are minimized (Finkel, 1990; Burmaster and Lehr, 1991; Paustenbach et al., 1991; Finley et al., 1994). We believe that the lower bound (5th percentile) of the HBAL PDF, which is theoretically protective of 95% of the population, should provide a reasonable and scientifically defensible soil action level.

BACKGROUND

The concept of setting a generic standard was originally utilized for drinking water supplies with the promulgation of maximum contaminant levels (MCLs) as part of the Safe Drinking Water Act (U.S. Department of Health, 1962; USEPA, 1991a). Traditionally, however, the USEPA and state regulators have set soil standards based on some form of site-specific assessment that considered site conditions such as soil chemistry, bioavailability, land use, and the potential for chemicals in soil to migrate to groundwater. Because MCLs are basically designed to be protective of one end use (human exposure via drinking water consumption), site-specific information is not as important.

While the concept of using the USEPA soil screening levels (SSLs) (USEPA, 1996a) runs somewhat contrary to the USEPA's historical position that cleanup levels for soil should be based on site-specific considerations and risk assessment (USEPA, 1990a), the SSL guidance provides a framework into which site information may be incorporated to develop site-specific SSLs with relatively little effort. The SSLs also provide a good example of an effective framework to screen chemicals in soil. An SSL is a chemical concentration in soil below which there is no health or environmental concern, provided all conditions associated with the SSLs are met (e.g., there are no ecological receptors of concern). SSLs do not trigger a response action or define "unacceptable" levels of contaminants in soil. Rather, the purpose of the SSLs is to provide a tool for (1) streamlining, standardizing, and accelerating the investigation and remediation process at contaminated properties, (2) adding consistency to decisions made about contaminated soil, and (3) focusing site-specific risk assessments on only the chemicals of greatest concern (USEPA, 1996a). All state and federal environmental regulatory agencies purport that the use of site-specific risk assessment has not been abandoned, rather that action levels are default levels to be used as "points of departure" from which an interested party may either conduct a risk assessment, declare that "no-action" is necessary, or remediate.

Because chromium is of concern at hundreds of state and federal hazardous waste sites (ATSDR, 1994), it is one metal for which action levels have been set by many states and USEPA regions. Although chromium may occur in nine valence states [Cr(-II) through Cr(VI)] (ATSDR, 1994), Cr(III) and Cr(VI) are typically the only environmentally stable and toxicologically important forms of chromium. Trivalent chromium is a naturally occurring and environmentally pervasive element that has low acute and chronic toxicity and has been recognized as an essential micronutrient for many years (Mertz, 1969). Hexavalent chromium rarely occurs naturally, is acutely toxic at high doses, and is recognized as a human respiratory carcinogen (USEPA, 1984; IARC, 1990; ATSDR, 1994). Additionally, Cr(VI) has been shown to be a skin irritant and to induce allergic contact dermatitis among persons who have been sensitized to it (Adams, 1990). Sensitization is thought to occur primarily from occupational exposures (Paustenbach et al., 1992; Nethercott et al., 1994) or from wet cement and construction materials during home remodeling projects.

As the valences of chromium have very different toxicological properties, it is appropriate to set separate action levels for each. Further, Cr(III) and Cr(VI) have very different fate and transport characteristics, that is, the leachability of Cr(VI) is enhanced under alkaline soil conditions, and the leachability of Cr(III) is enhanced under strongly acidic soil conditions (Saleh et al., 1989). For Cr(VI), soil suspension to ambient air by wind or mechanical friction (e.g., traffic on unpaved soil) is a significant process, whereas it is not a significant concern for Cr(III) because of its low inhalation toxicity.

SURVEY OF STATE AND FEDERAL SOIL ACTION LEVELS

In January and February of 1996, environmental departments from each of the 50 states as well as the 10 USEPA regions were contacted to determine if they have developed and utilize action levels for chromium in soil, or are currently developing such criteria. The following information was determined from the survey: whether the levels have been developed or adopted from another source; which species of chromium are regulated; land use action level (i.e., residential, industrial, etc.); the basis for the standard (i.e., route of exposure, protection of groundwater, etc); and if site-specific risk assessments may be used to develop cleanup values *in lieu* of the non-site-specific levels. This information was obtained from verbal communication and regulatory guidance documents and recorded (Table 1). The state values are generically termed "action levels" for the purpose of this discussion as the terms used by the states are not consistent. It should be noted that some of the action levels used in this evaluation may not be finalized values (Table 1).

Survey Results

Of the 10 EPA regions, two have developed their own guidance (Regions III and IX), and a third (Region VII) has compiled criteria from several other sources (Table 1). Region III developed the Risk Based Concentration Limits (RBCLs), and Region IX developed the Preliminary Remediation Goals (PRGs). Region VII guidance includes action levels from Region III, Missouri and Kansas. The most common action levels used by the regions are the USEPA SSLs. Regions I, II, III, V, VI, and X use the SSLs, either as their sole guidance or in addition to their own guidance. The Region III RBCLs are also used by Regions VII and X. Region VII reported additionally using the RCRA Subpart S guidance. Region II commonly develops acceptable cleanup levels by backcalculating soil levels from Federal Maximum Contaminant Levels (MCLs) for protection of groundwater; however, they may also use USEPA SSLs. Regions VIII and IV are unique in that they have developed their own guidance for modifying the RBCLs and SSLs. Both regions recommend using Region III RBCLs or USEPA SSLs and multiplying values for non-carcinogens by a factor of 0.1 to protect the potential for additive chemical toxicity at sites with multiple chemicals in soil.

Currently, 42 states apply some type of action level to chromium in soil, although 4 of these simply use a leaching test to determine the potential for chromium to impact groundwater (IA, NE, SD, and NH) (Table 1). Eight states (AL, AR, MD, ND, OK, RI, UT, and WY) rely solely on site-specific risk assessment to derive soil cleanup levels. At the time of this survey, however, two of these eight states (AR and MD) were in the process of developing action levels that would bring the total number of states with soil remediation action levels to 44 sometime in 1997.

Of the 42 states utilizing action levels for chromium in soil, 38 base their levels on a health risk assessment model, and 26 have developed their own soil action levels. Twelve states simply adopted federal or regional EPA guidance/approaches. The most commonly state-adopted values are the Region III RBCLs (USEPA, 1995a) and the USEPA Subpart S action levels. Also frequently utilized by states are the USEPA SSLs (USEPA, 1996a) and the Region IX PRGs (USEPA, 1995b).

Despite the general consistency in approaches, the action levels generated by the environmental agencies vary widely (Table 2). To further assess the source of variance, soil action levels were grouped for further analysis by land use and

TABLE 1
STATE AND FEDERAL ACTION LEVELS FOR CHROMIUM

State/Region	Action Levels and/or Guidance Utilized	Basis and Comments
ALABAMA (334)271-7738	Soil cleanup criteria are based on site-specific risk assessments.	
ALASKA (907)269-7549	*Draft Oil and Hazardous Substances Pollution Control - Regulations*, December 18, 1996 - Updated October 25, 1996 Residential: Total Cr(mg/kg) — Arctic Zone — Under 40" Zone — Over 40" Zone ingestion — 680 — 510 — 410 groundwater — - — 25 — 22 mitigation	The arctic zone is an area with continuous permafrost, and the 40 inch zones refer to areas receiving over and under 40 inches of precipitation per year. These values are based on the ingestion pathway.
ARIZONA (602)207-4180	*Interim Soil Remediation Standards*, September 22, 1995 May remediate to one of the following levels: • Background, • Site-specific risk assessment cleanup levels, or • Health-based guidance levels (HBGLs): (mg/kg) — Residential — Non-Residential Cr(III) — 77,000 — 100,000 Cr(VI) — 30 — 64 Total Cr — 2,100 — 4,500	HBGLs are based on the ingestion pathway. Anything less protective than 10^{-4} will be restricted to non-residential use. Total chromium is based on a Cr(VI)/Cr(III) ratio of 1/6.
ARKANSAS (501)682-0874	*Remedial Actions Trust Fund Act 125 of 1995*- DRAFT (Voluntary Cleanup Program), September 1 Under the Arkansas Brownfields Program, the state is considering the following three tiered system: • Background • Media-specific statewide standards (not yet developed) • Site-specific risk assessment cleanup levels	
CALIFORNIA (916)255-2047	*Preliminary Endangerment Assessment (PEA) Guidance Manual*, 1994 (mg/kg) — Residential Cr (VI) — 0.2 Cr(III) — 70,000	California considers Cr(VI) and compounds to be oral and inhalation carcinogens; these PEAs are based on the ingestion pathway. The PEA guidance provides risk and hazard equations from which these action levels were calculated.

600

TABLE 1 (Cont.)
STATE AND FEDERAL ACTION LEVELS FOR CHROMIUM

State/Region	Action Levels and/or Guidance Utilized	Basis and Comments
COLORADO (303)692-3362	Use: • Subpart S guidance (40 CFR, Part 264), July 27, 1990 Cr(VI) action level: 400 mg/kg • TCLP test for groundwater	Subpart S criteria are based on the ingestion pathway. Colorado is in the process of developing guidance.
CONNECTICUT (203)424-3705	*Remediation Standard Regulations*, December 13, 1995 May screen using background or the following Direct Exposure Criteria: (mg/kg) Residential Industrial/Commercial Cr(III) 3,900 51,000 Cr(VI) 100 100	Connecticut direct exposure criteria are based on the ingestion and dermal contact pathways.
DELAWARE (302)323-4540	*Hazardous Substance Cleanup Program*, June 17, 1994 - Updated October, 1995 Reporting Levels: (mg/kg) Residential Industrial Subsurface Cr(VI) 390 10,000 19	Residential and industrial reporting levels are based on the ingestion pathway. The "subsurface" reporting level is based on groundwater protection. Delaware also has a Brownfields program (November 1995) and a Voluntary Cleanup Program (February 1995).
FLORIDA (904)921-9983	*Soil Cleanup Goals for Florida*, September 27, 1995 Soil Cleanup Goals: (mg/kg) Residential Industrial Cr(III) 660,000 540,000 Cr(VI) 290 430	Florida soil cleanup goals are based on the inhalation, ingestion, and dermal contact pathways.
GEORGIA (404)657-8600	*Hazardous Site Response*, June 16, 1993 Soil Criteria: Type 1 Residential target concentration of 100 mg/kg total Cr. Type 2 Residential site-specific risk assessment. Type 3 Non-Residential target concentration of 100 mg/kg total Cr. Type 4 Non-Residential site-specific risk assessment. Type 5 Any type other than 1-4, requiring engineered barriers and instituted controls.	Georgia's soil criteria are based the ingestion pathway.

TABLE 1 (Cont.)
STATE AND FEDERAL ACTION LEVELS FOR CHROMIUM

State/Region	Action Levels and/or Guidance Utilized	Basis and Comments
HAWAII (808)586-4226	Use: • EPA Region IX *Preliminary Remediation Goals* (PRGs), August 1, 1996 Groundwater Protection (mg/kg) Residential Industrial 20 DAF 1 DAF Total Cr 210 450 38 2 Cr(VI) 30 64 38 2	Region IX PRGs are based on the inhalation pathway and assume that total Cr is a 1/6 ratio of Cr(VI)/Cr(III).
IDAHO (208)373-0502	Use: • Subpart S guidance (40 CFR, Part 264), July 27, 1990 Cr(VI) action level: 400 mg/kg • TCLP test for groundwater	Subpart S criteria are based on the ingestion pathway.

ILLINOIS
(217)308-3159
	Tiered Approach to Cleanup Objectives Guidance Document, January 1, 1996- Updated April 1997
	Can use background to screen or one of the following (as appropriate):
	<u>Tier 1</u> -
	• Route-specific baseline contaminant cleanup objectives:
	
	                                      Industrial/ Construction
	                Residential Commercial Worker
	(mg/kg) ingestion inhalation ingestion inhalation ingestion inhalation
	Cr Total/Cr(VI) 390 270 10,000 420 4,100 8,800
	Cr(III) 78,000 - 1,000,000 - 330,000 -
	
	Groundwater protective cleanup objectives. May choose leachability based or pH based objectives:
	• Groundwater Protection (class I groundwater only) (mg/kg):
	  pH 4.5+ 4.75+ 5.25+ 5.75+ 6.25+ 6.65+ 6.9+ 7.25+ 7.75- 8.0
	  Cr(VI) 70 62 54 46 40 38 36 32 28
	<u>Tier 2</u> - Site-specific risk assessment used to develop cleanup levels. This method allows for engineered barriers and institutional controls described in these guidelines.
	<u>Tier 3</u> - A simple or complex site assessment allowing alternative parameters and/or factors not available under Tier 1 and 2.
		Ingestion cleanup levels are based on a hazard index of 1, inhalation cleanup levels are based on an excess cancer risk of 10^{-6}. Leachability based groundwater protective cleanup objectives are based on the TCLP. Tiers 2 and 3 require site-specific risk assessments.

TABLE 1 (Cont.)

STATE AND FEDERAL ACTION LEVELS FOR CHROMIUM

State/Region	Action Levels and/or Guidance Utilized	Basis and Comments
INDIANA (317)233-6454	*Voluntary Remediation Program*, October 1995 Tier 1 - Background Tier 2 - Cleanup goals: Residential and (mg/kg) Non-residential Residential Non-residential Subsurface Cr(III) 10,000 10,000 10,000 Cr(VI) 10,000 1,350 7,3000 Tier 3 - Any other unique situation.	Indiana cleanup goals are based on the inhalation pathway.
IOWA (515)281-7040	Chapter 133: Department Rule, July 20, 1989 Groundwater lifetime adult action level for total chromium: 0.1 mg/L	Iowa does not have cleanup criteria specific to soil.
KANSAS (913)296-1675	*Interim Remedial Guidelines for Contaminated Soils*, August 1995 (mg/kg) Residential Non-residential RCRA 20 Rule Total Cr 120 1,700 100	Total chromium concentrations are based on the inhalation, ingestion, and dermal contact pathways. The RCRA 20 Rule action level is based on the protection of groundwater. A TCLP test must be performed when this level is exceeded.
KENTUCKY (502)564-6116	*Remedial Options Guidance*, August 15, 1995 Use human health guidance screening level of 30 mg/kg for Cr(VI). Have a three-tiered remedial options program: Tier 1 - Background or removal of hazardous substance, pollutant, or contaminant. Tier 2 - Site-specific risk assessment. Tier 3 - Risk assessment with no action.	Kentucky's screening level is based on the inhalation, ingestion and dermal contact pathways.
LOUISIANA (504)765-0249	*DRAFT - Proposed Approach for Implementing a Louisiana*, August 9, 1995 Department of Environmental Quality Risk Based Corrective Action Program Corrective Option 1: Use non-industrial and industrial corrective action levels for total Cr & Cr(VI) of 100 mg/kg. Corrective Option 2: Conduct a baseline risk assessment to derive cleanup levels where excess cancer risk is no greater than 10^{-4}. Corrective Option 3: When corrective Option 2 can not be achieved, levels achievable by best demonstrated available technology can be developed.	Louisiana corrective action levels are based on groundwater protection.
MAINE (207)287-6455	*Guidance Manual for Human Health Risk Assessment at Hazardous Sites*, June 1994 Also have draft guidelines: (mg/kg) Residential Trespasser Industrial Cr(VI) 950 5,350 10,000	This manual provides state guidance for developing site-specific cleanup levels. Maine draft soil criteria are based on the ingestion pathway.

TABLE 1 (Cont.)

STATE AND FEDERAL ACTION LEVELS FOR CHROMIUM

State/Region	Action Levels and/or Guidance Utilized	Basis and Comments
MARYLAND (410)631-3437	Maryland is currently developing a Voluntary Cleanup Program. Cleanup standards will probably be available in early 1998.	
MASSACHUSETTS (617)556-1160	*The Massachusetts Contingency Plan*, February 2, 1995 Method 1- Soil Standards This method is to be used when direct contact and ground water are of concern. (mg/kg) Soil Category 1 Soil Category 2 Soil Category 3 Total Cr 1,000 2,500 5,000 Cr(III) 1,000 2,500 5,000 Cr(VI) 200 600 1,000 Method 2- Soil Standards Using Method 1 soil standards, method 2 may supplement method 1 with certain site-specific parameters. Method 3- Upper Concentration Limits (UCLs). UCLs are used to determine if media concentrations pose a significant risk to human health. The UCLs for total Cr, Cr(III), and Cr(VI) are all 10,000 mg/kg.	All concentrations are based on the ingestion and dermal contact pathways. Soil categories 1, 2, and 3 respectively represent land uses similar to residential, commercial, and industrial.
MICHIGAN (517)373-4800	*Interim Specific Remedial Action Plan*, June 5, 1995 Generic residential cleanup criteria: (mg/kg) Cr(III) 630,000 Cr(VI) 2,000 Generic industrial and commercial cleanup criteria: Commercial Subcategories Groundwater Protection (mg/kg) Industrial III IV 20 X DW 20 X GSI Cr(III) 1,000,000 1,000,000 1,000,000 2 1.5 Cr(VI) 22,000 30,000 71,000 2 0.15	All concentrations are based on the ingestion and dermal contact pathways. There are no specific cleanup criteria for subcategories I and II, as Subcategory I most likely warrants the use of residential criteria, and subcategory II most likely falls under the industrial category. The groundwater protection action levels are based on 20 times the drinking water values and 20 times the groundwater surface interface values, both listed in this guidance.
MINNESOTA (612)296-7358	*Draft Interim Screening Values*, April, 26, 1996 (mg/kg) Residential Groundwater Protection Ecological Soil Screening Cr(III) 24,000 246,376 - Cr(VI) 126 62 0.4	Screening values are based on the inhalation, ingestion, and dermal contact pathways. The ecological soil screening level is based on a hazard quotient of 0.

TABLE 1 (Cont.)
STATE AND FEDERAL ACTION LEVELS FOR CHROMIUM

State/Region	Action Levels and/or Guidance Utilized	Basis and Comments
MISSISSIPPI (601)961-5072	*Guidance for Remediation of Uncontrolled Hazardous Substance Sites in Mississippi,* September 1990 - • USEPA Soil Screening Levels (SSLs), May 1996 Groundwater Protection: (mg/kg) Inhalation Ingestion 20 X DAF 1 X DAF Cr(III) - 78,000 38 2 Cr(VI) 270 390 38 2	SSL groundwater protection values are based on 1 and 20 times the dilution-attenuation factor.
MISSOURI (314)751-3176	*How Clean is Clean? - Uniform cleanup standards for contaminated sites in Missouri,* April 1995 Tier 1- Uniform cleanup standards Any use soil levels (ASL): • 5,600 mg/kg total Cr concentration where Cr(VI) is unlikely, and • 280 mg/kg total Cr concentration where Cr(VI) is likely. Tier 2- Alternative cleanup standards (ACSs): Health-based ACSs may be developed if site exceeds background.	ASLs are based on the ingestion pathway. Cr(VI) is considered to be likely if it is documented above 4 mg/kg. Missouri currently has an updated version of *How Clean is Clean?* under review; it is likely that the ASLs will change.
MONTANA (406)444-0478	May use the following: • site-specific risk assessment, or • USEPA Soil Screening Levels (SSLs), May 1996 Groundwater Protection: (mg/kg) Inhalation Ingestion 20 X DAF 1 X DAF Cr(III) - 78,000 38 2 Cr(VI) 270 390 38 2	SSL groundwater protection values are based on 1 and 20 times the dilution-attenuation factor.
NEBRASKA (402)471-4210	*Title 128 - Rule and Regulations Governing Hazardous Waste Management in Nebraska,* September 20, 1995 Regulatory Level: 5 mg/L	Nebraska does not have cleanup criteria specific to soil, but use the TCLP to determine remediation standards.
NEVADA (707)687-4670	*Contaminated Soil and Groundwater Remediation Policy,* June 25, 1992 Use most conservative of all that apply: • If groundwater is of concern, use the TCLP test. • If incidental soil ingestion or dermal contact is of concern use: Subpart S guidance (40 CFR, Part 264), July 27, 1990: Cr(VI) action level: 400 mg/kg • If surface water is a factor, use Nevada Water Quality Standards (NAC 445.117).	The Subpart S Cr(VI) action level is based on the ingestion pathway.

TABLE 1 (Cont.)

STATE AND FEDERAL ACTION LEVELS FOR CHROMIUM

State/Region	Action Levels and/or Guidance Utilized				Basis and Comments
NEW HAMPSHIRE (603)271-2942	Use TCLP test as a screening guidance.				New Hampshire does not have cleanup criteria specific to soil.
NEW JERSEY (609)292-8294	*Draft Site Remediation Program, September 1995*				Residential and industrial concentrations for Cr(III) are based the ingestion pathway, and for Cr(VI) are based on the inhalation pathway. The all soil concentration for Cr(III) is an ecological screening level, and the all soil concentration for Cr(VI) is based on Allergic Contact Dermatitis. This guidance is currently under review.
	(mg/kg)	Residential	Industrial	All Soil	Vegetation & Phytotoxicity	
	Cr(III)	56,000	1,000,000	200	200	
	Cr(VI)	130	190	15		
NEW MEXICO (505)827-1558	Use: •EPA Region IX *Preliminary Remediation Goals,* (PRGs) August 1, 1996					Region IX PRGs are based on the inhalation pathway and assume that total Cr is a 1/6 ratio of Cr(VI)/Cr(III).
				Groundwater Protection		
	(mg/kg)	Residential	Industrial	20 DAF	1 DAF	
	Total Cr	210	450	38	2	
	Cr(VI)	30	64	38	2	
NEW YORK (518)457-0337	DRAFT *Determination of Soil Cleanup Objectives and Cleanup Levels TAGM (technical and administrative guidance memorandum),* January 24, 1994 Recommended soil cleanup objectives: •50 mg/kg total Cr, or •site background. May cleanup to: •Predisposal concentration, or •Site-specific cleanup levels.					New York's cleanup objectives are based on background levels, and are currently being raised from 10 mg/kg to 50 mg/kg.
NORTH CAROLINA (919)773-2178	*Guidelines for Responsible Party Voluntary, Site Remediation Action, Inactive Hazardous Sites Program,* October 1996 Soil Remediation Goals:					North Carolina used a hazard quotient of 0.2 for non-carcinogens.
	(mg/kg)					
	Cr(III)	15,600				
	Cr(VI)	78				

606

TABLE 1 (Cont.)
STATE AND FEDERAL ACTION LEVELS FOR CHROMIUM

State/Region	Action Levels and/or Guidance Utilized	Basis and Comments
NORTH DAKOTA (701)328-5168	Soil cleanup criteria are based on site-specific risk assessments.	
OHIO (614)664-2295	*DRAFT Voluntary Action Plan*, October 31, 1995 - Updated June 1996. Maximum allowable single chemical cleanup levels: (mg/kg) / Residential / Commercial / Industrial / Groundwater Cr(III) 310 12000 8,400 - Cr(VI) 220 1,700 1,700 - Cr Total - - - 350	Ohio action levels are based on the ingestion pathway.
OKLAHOMA (402)664-2295	Soil cleanup criteria are based on site-specific risk assessments.	
OREGON (503)229-6764	Environmental Cleanup Manual, June 1994 May cleanup to one of the following: •Background, •Simple Site Cleanup: Leachability is not to exceed 10 µg/L total Cr and soil concentrations are not to exceed 10 mg/kg total Cr. •Complex Site Cleanup: Requires site-specific risk assessment. •Maximum allowable soil concentrations: (mg/kg) / Residential / Industrial / Groundwater Cr total 1,000 1,500 10	Maximum allowable soil concentrations are based on the ingestion pathway.
PENNSYLVANIA (717)657-4585	*Pennsylvania Land Recycling Program* (proposed), August 1996 - Updated November 1996 May cleanup to one of the following: •Background, •Medium-specific, statewide health cleanup objectives (not yet determined), or •Site-specific cleanup levels. Statewide Health Standards: (mg/kg) / Residential / Non-Residential / Soil to Groundwater Total Cr 1,000 1,000,000 10 Cr(VI) 300 5,000 -	Pennsylvania statewide health standards are based on the ingestion pathway.
RHODE ISLAND (401)277-2234	Use background to determine if concentration is reportable and determine cleanup concentrations by site-specific risk assessments.	

TABLE 1 (Cont.)
STATE AND FEDERAL ACTION LEVELS FOR CHROMIUM

State/Region	Action Levels and/or Guidance Utilized	Basis and Comments
SOUTH CAROLINA (803)734-5000	Use: EPA Region III *Risk-Based Concentration Limits* (RBCLs), March, 1997 Transfer from soil to: (mg/kg) Industrial Residential Groundwater Air Cr(III) 1,000,000 78,000 19 140 Cr(VI) 10,000 390 - -	Industrial and residential RBCLs for chromium are based on the ingestion pathway.
SOUTH DAKOTA (605)773-3296	Presently use TCLP test to develop a cleanup level.	South Dakota does not have cleanup criteria specific to soil.
TENNESSEE (615)532-0900	May cleanup to one of the following: • EPA Region III *Risk-Based Concentration Limits* (RBCLs), March, 1997 Transfer from soil to: (mg/kg) Industrial Residential Groundwater Air Cr(III) 1,000,000 78,000 19 140 Cr(VI) 10,000 390 • 2 times the average background as a cleanup level.	Industrial and residential RBCLs for chromium are based on the ingestion pathway.
TEXAS (512) 239-2343	*Texas Risk Reduction Rule*, 1993 Can remediate to one of the following: Standard 1: Background Standard 2: Media-specific statewide standards (mg/kg): Residential Industrial Groundwater Protection Cr(VI) 391 5,110 10 Total Cr 391 5,110 10 Standard 3: Site-specific cleanup levels.	Texas residential and industrial standards are based on the ingestion pathway.
UTAH (801)536-4170	Soil cleanup criteria are based on site-specific risk assessments.	Utah has a newly developed voluntary cleanup program.
VERMONT (802)241-3888	May use one of the following as guidance; • EPA Region III *Risk-Based Concentration Limits* (RBCLs), March, 1997 Transfer from soil to: (mg/kg) Industrial Residential Groundwater Air Cr(III) 1,000,000 78,000 19 140 Cr(VI) 10,000 390 • Federal drinking water standards.	Industrial and residential RBCLs for chromium are based on the ingestion pathway.

TABLE 1 (Cont.)
STATE AND FEDERAL ACTION LEVELS FOR CHROMIUM

State/Region	Action Levels and/or Guidance Utilized	Basis and Comments
VIRGINIA (804)698-4118	Generally remediate to: • Background, • Site-specific risk assessment cleanup levels, or • EPA Region III *Risk-Based Concentration Limits (RBCLs)*, March, 1997 Transfer from soil to: (mg/kg) — Industrial — Residential — Groundwater — Air Cr(III) — 1,000,000 — 78,000 — 19 — 140 Cr(VI) — 10,000 — 390 — - — -	Industrial and residential RBCLs for chromium are based on the ingestion pathway. Virginia uses Risk Exposure Analysis Model System (REAMS) to calculate cleanup levels.

WASHINGTON (360)407-7188	*Model Toxics Control Act: Cleanup Levels and Risk Calculations*, December 1995
	
	(mg/kg) — Residential — Industrial
	• Method A
	  Cr Total — 100 — 500
	• Method B
	  Cr(III) — 80,000 — -
	  Cr(VI) — 400 — -
	• Method C
	  Cr(III) — 320,000 — 3,500,000
	  Cr(VI) — 1,600 — 17,500
		Method A is based on the inhalation pathway, and Methods B and C are based on the ingestion pathway.

WEST VIRGINIA (304)558-6350	Use: • Subpart S guidance (40 CFR, Part 264), July 27, 1990 Cr(VI) action level: 400 mg/kg • TCLP test for groundwater	Subpart S guidance is based on the ingestion pathway.
WISCONSIN (608)266-5425	*Investigation and Remediation of Environmental Contamination*, December, 1995 May remediate to: • Site-specific cleanup level, or • Residual contaminant levels: (mg/kg) — Non-Industrial — Industrial Cr(III) — 16,000 Cr(VI) — 14 — 200	Cr(VI) levels are based on the inhalation pathway, and an acceptable cancer risk of 10^{-7} to protect for multiple chemicals
WYOMING (307)777-7740	May remediate to: • Background, or • Site-specific cleanup levels.	

TABLE 1 (Cont.)
STATE AND FEDERAL ACTION LEVELS FOR CHROMIUM

State/Region	Action Levels and/or Guidance Utilized	Basis and Comments
Region I (617)223-5541	Use: • USEPA Soil Screening Levels (SSLs), May, 1996 **(mg/kg)** — **Inhalation** — **Ingestion** — **Groundwater Protection: 20 X DAF / 1 X DAF** Cr(III) — - — 78,000 — 38 — 2 Cr(VI) — 270 — 390 — 38 — 2	SSL groundwater protection values are based on 1 and 20 times the dilution-attenuation factor.
Region II (212)637-4311	May use either of the following, however the first is the most common method: • Back-calculate acceptable soil levels from the Federal MCLs; May 1995 - *Drinking water Regulations and Health Advisories* (Proposed) **Total Cr** — **MCL** — **MCLG** (mg/L) — 0.1 — 0.1 • USEPA Soil Screening Levels (SSLs), May 1996 **(mg/kg)** — **Inhalation** — **Ingestion** — **Groundwater Protection: 20 X DAF / 1 X DAF** Cr(III) — - — 78,000 — 38 — 2 Cr(VI) — 270 — 390 — 38 — 2	The MCLs are based on ingestion of groundwater; soil criteria should be protective of groundwater. SSL groundwater protection values are based on 1 and 20 times the dilution-attenuation factor.
Region III (215)566-3319	Can use either: • EPA Region III *Risk-Based Concentration Limits* (RBCLs), March, 1997 Transfer from soil to: **(mg/kg)** — **Industrial** — **Residential** — **Groundwater** — **Air** Cr(III) — 1,000,000 — 78,000 — - — - Cr(VI) — 10,000 — 390 — 19 — 140 or • USEPA Soil Screening Levels (SSLs), May 1996 **(mg/kg)** — **Inhalation** — **Ingestion** — **Groundwater Protection: 20 X DAF / 1 X DAF** Cr(III) — - — 78,000 — 38 — 2 Cr(VI) — 270 — 390 — 38 — 2	Industrial and residential RBCLs for chromium are based on the ingestion pathway. Region III updates the RBC table semi-annually. SSL groundwater protection values are based on 1 and 20 times the dilution-attenuation factor.

TABLE 1 (Cont.)
STATE AND FEDERAL ACTION LEVELS FOR CHROMIUM

State/Region	Action Levels and/or Guidance Utilized	Basis and Comments
Region IV (404)347-6143	*Supplemental Guidance to RAGS: Region 4 Bulletin Human Health Risk Assessment* (Interim) - November, 1995	This guidance recommends that the USEPA SSLs be divided by a factor of 10 to develop Region IV action levels.
Region V (312) 886-4904	Use: • USEPA Soil Screening Levels (SSLs), May 1996 Groundwater Protection: (mg/kg) / Inhalation / Ingestion / 20 X DAF / 1 X DAF Cr(III) / - / 78,000 / 38 / 2 Cr(VI) / 270 / 390 / 38 / 2	SSL groundwater protection values are based on 1 and 20 times the dilution-attenuation factor.
Region VI (214)665-2270	May use either of the following: Use: • Subpart S guidance (40 CFR, Part 264), July 27, 1990 Cr(VI) action level: 400 mg/kg • EPA Region III *Risk-Based Concentration Limits* (RBCLs), March, 1997 Transfer from soil to: (mg/kg) / Industrial / Residential / Groundwater / Air Cr(III) / 1,000,000 / 78,000 / - / 140 Cr(VI) / 10,000 / 390 / 19 / -	Subpart S guidance and the Industrial and residential RBCLs for chromium are based on the ingestion pathway.
Region VII (913) 551-7821	June 1996 - *EPA Region VII Screening Table* This table provides the following soil screening information: • Missouri ASLs: 50,000 mg/kg Cr(III), 280 mg/kg Cr(VI); • Kansas IRGs (total Cr): residential - 120 mg/kg, nonresidential - 1,700 mg/kg; • Superfund Chemical Data Matrix: 580,000 mg/kg Cr(III), 2,900 mg/kg Cr(VI)/total Cr; • EPA Region III *Risk-Based Concentration Limits* (RBCLs), March, 1997 Transfer from soil to: (mg/kg) / Industrial / Residential / Groundwater / Air Cr(III) / 1,000,000 / 78,000 / - / 140 Cr(VI) / 10,000 / 390 / 19 / -	Industrial and residential RBCLs for chromium are based on the ingestion pathway.
Region VIII (303)312-6210	*Region 8 Superfund Technical Guidance, Evaluating and Identifying Contaminants of Concern for Human Health,* September 1994	This guidance recommends dividing a RAGS Part B-derived action level (*e.g.*, Region III RBCLs or USEPA SSLs) by a factor of 10 to derive Region XIII action levels. This method is executed to protect against additive effects of chemicals.

TABLE 1 (Cont.)
STATE AND FEDERAL ACTION LEVELS FOR CHROMIUM

State/Region	Action Levels and/or Guidance Utilized	Basis and Comments
Region IX	EPA Region IX *Preliminary Remediation Goals*, (PRGs) August 1, 1996 Groundwater Protection (mg/kg) — Residential — Industrial — 20 DAF — 1 DAF Total Cr — 210 — 450 — 38 — 2 Cr(VI) — 30 — 64 — 38 — 2	Region IX tables are updated annually. Region IX PRGs are based on inhalation of suspended particulates and assume that total Cr is a 1/6 ratio of Cr(VI)/Cr(III).
Region X (206)553-0125	May use either of the following: • EPA Region III *Risk-Based Concentration Limits* (RBCLs), March, 1997 Transfer from soil to: (mg/kg) — Industrial — Residential — Groundwater — Air Cr(III) — 1,000,000 — 78,000 — - — - Cr(VI) — 10,000 — 390 — 19 — 140 • USEPA Soil Screening Levels (SSLs), May 1996 Groundwater Protection: (mg/kg) — Inhalation — Ingestion — 20 X DAF — 1 X DAF Cr(III) — - — 78,000 — 38 — 2 Cr(VI) — 270 — 390 — 38 — 2	Industrial and residential RBCLs for chromium are based on the ingestion pathway. SSL groundwater protection values are based on 1 and 20 times the dilution-attenuation factor.

TABLE 2
Summary of Human Health State Action Levels for Chromium: Differentiated by Land-Use and Species

	Minimum (mg/kg)	Maximum (mg/kg)	Mode (mg/kg)
Residential Cr(III)	200[a]	660,000[b]	78,000[c]
Residential Cr(VI)	0.2[d]	2,000[e]	390[f]
Non-residential Cr(III)	2,500[g]	>1,000,000[h]	1,000,000[i]
Non-residential Cr(VI)	64[j]	71,000[k]	10,000[l]

[a] NJ
[b] FL
[c] IL, MS, MT, SC, TN, VA, and VT (AZ = 77,000 mg/kg)
[d] CA
[e] MI
[f] DE, IL, MS, MT, SC, TN, VA, and VT (TX = 391 mg/kg)
[g] MA
[h] WA
[i] IL, MI, NJ, SC, TN, VA, and VT (WA = >1,000,000 mg/kg)
[j] HI, NM, (both use Region IX PRGs), and AZ.
[k] DE, IL, IN, ME, SC, TN, VA, and VT

valence designations. Most states differentiate chromium soil action levels for residential* and non-residential** land use scenarios, and specify levels for Cr(III) and Cr(VI) valence species separately. The ranges of action levels for both land uses and species of chromium are presented in Table 2. Trivalent chromium action levels for a residential scenario have a mode of 78,000 mg/kg, which is the Region III RBCL. Overall, the action levels are generally above 10,000 mg/kg, but five states have action levels at or below 6000 mg/kg (CT, MO, MA for category 1 soils, NJ's for preventing phytotoxicity, and OH using a probabilistic [Monte Carlo] analysis). For a non-residential land use, the action levels for Cr(III) are much higher, with a mode of 1,000,000 mg/kg (unity). Of the 14 states with non-residential Cr(III) action levels, only two developed values lower than 10,000 mg/kg (MA for category 1 and 3 soils and one option developed by Ohio).

As one would expect, Cr(VI) action levels are significantly lower than those for Cr(III). The mode for Cr(VI) residential action levels is the Region III RBCL of 390 mg/kg (Table 2), which is protective of soil-ingestion exposures for children at residential sites. Although 9 states use this value, there are 24 states that use lower action levels, and only 9 that use higher values. Other common residential Cr(VI) action levels are the Region IX PRG (30 mg/kg) and the Region III RBCL

* Residential land uses include those specified as "residential" and those with only one land use presented.
** Non-residential land uses include the following scenarios: industrial, commercial, construction and trespasser.

for suspension of soil to air (140 mg/kg). The mode of non-residential Cr(VI) action levels is 10,000 mg/kg. Those states using Region IX values have the lowest non-residential action levels of 64 mg/kg.

Total chromium action levels have been set by 14 states (AK, AZ, GA, HI, AL, IL, KS, LA, MA, NM, NY, OR, PA, TX, and WA). Both the residential and non-residential mode for total chromium action levels are 100 mg/kg. It should be noted that the USEPA soil digestion/analysis method for total chromium (Method 6010; USEPA, 1990b) uses a nitric acid digestion, and thus any Cr(VI) in the sample would be reduced to Cr(III) yielding results only in terms of total chromium. Cr(VI) may be analyzed in soil using a separate method (Method 3060A/7196A; USEPA, 1995c; Vitale *et al.*, 1993), and the Cr(III) concentration can then be calculated by subtracting the Cr(VI) concentration from the total chromium concentration. Because Cr(III) prevails in the environment, measurements of "total chromium" are frequently considered to be all Cr(III); however, levels for "total chromium" are more often protective of Cr(VI) than Cr(III).

Sources of Variation

Despite the differentiation of standards by land use and by chromium species, considerable variance in chromium action levels exists. To further understand the source of variance, the state criteria were additionally divided by route of exposure (e.g., soil ingestion or inhalation) (Table 3). Seventeen states (AZ, FL, IL, HI, KS, KY, MN, MS, MT, NJ, NM, SC, TN, VT, VA, WA, and WI) use levels that include soil suspension modeling and the inhalation pathway of exposure, and 28 states (AK, CA, CO, CT, DE, GE, ID, IL, MA, ME, MI, MO, MS, MT, NV, NJ, NC, OH, OR, PA, SC, TN, TX, VA, VT, WV, and WA) use levels based on the soil ingestion pathway. It should be noted that some states, most of which utilize either the Region III RBCLs or the USEPA SSLs, have both an inhalation-based level, developed using soil suspension modeling, and an ingestion-based value. Likewise, some states base their action levels on something other than ingestion or inhalation (i.e., naturally occurring background, protection of groundwater). Generally, Cr(VI) state action levels protective of the inhalation pathway are lower than those of states that only consider soil ingestion. It should be noted that the soil suspension modeling component of the inhalation-based level adds significant complexity to the risk assessment model, and thus most states evaluate only soil ingestion to simplify the analysis. Further, soil suspension modeling results are very sensitive to site-specific parameters such as traffic conditions (vehicle weight, number of wheels, distance traveled), weather conditions (rain and wind speed), land cover (vegetation, pavement), and soil parameters (particle size) (Cowherd *et al.*, 1985). Thus, generic standards that must consider all of these variables are very uncertain.

TABLE 3
**Summary of State Action Levels for Chromium:
Differentiated by Land Use, Species and Pathway of Exposure**

	Minimum (mg/kg)	Maximum (mg/kg)	Mode (mg/kg)	Number of states
Based on soil ingestion				
Residential Cr(III)	310[a]	630,000[b]	78,000[c]	17
Residential Cr(VI)	0.2[d]	2,000[b]	390[e]	24
Non-residential Cr(III)	2,500[f]	>1,000,000[g]	1,000,000[h]	12
Non-residential Cr(VI)	100[i]	71,000[b]	10,000[j]	15
Based on soil suspension to air and inhalation				
Residential Cr(III)	16,000[k]	660,000[l]		
Residential Cr(VI)	14[k]	290[l]	30; 140[m]	15
Non-residential Cr(III)	100,000[n]	540,000[l]	—	2
Non-residential Cr(VI)	64[o]	8,800[p]	64[o]	7
Based on groundwater protection				
Cr(III)	1.5[b]	246,376[q]	—	2
Cr(VI)	0.15[b]	100[r]	19[s]	14
Cr(total)	2.0[t]	350[a]	2; 38[j]	11

Note: No mode for this scenario.

[a] OH.
[b] MI.
[c] IL, MS, MT, SC, TN, VA, and VT.
[d] CA.
[e] DE, IL, MS, MT, SC, TN, VA, and VT (TX: 391 mg/kg).
[f] MA.
[g] WA.
[h] IL, MI, NJ, SC, TN, VA, and VT (WA: >1,000,000 mg/kg).
[i] CT.
[j] DE, IL, IN, ME, SC, TN, VA, and VT.
[k] WI.
[l] FL.
[m] 30 mg/kg: AZ, HI, KY and NM; 140 mg/kg: VA, VT, SC, and TN.
[n] AZ.
[o] HI, NM, and AZ.
[p] IL.
[q] MN.
[r] LA.
[s] DE, SC, TN, VA, and VT.
[t] HI, NM, MS, and MT.

Eighteen states (AK, DE, IL, KS, MI, OH, OR, PA, TX, MN, HI, NM, MI, MT, SC, TN, VT, and VA) utilize chromium soil action levels protective of groundwater in addition to soil ingestion and inhalation-based action levels. Louisiana has only a groundwater protection action level, bringing the total of states with groundwater protection action levels to 19. The premise of groundwater protection action levels is that there is a potential for chromium in soil to leach to groundwater drinking water sources. In general, these values are more conservative than ingestion and inhalation-based action levels; however, some agencies stated that groundwater action levels can be used as levels that trigger site-specific TCLP-type tests. Two states developed groundwater protective action levels for Cr(III) with considerable variance in values; Michigan has two: 2.0 mg/kg and 1.5 mg/kg, and Minnesota has one: 246,376 mg/kg. Twelve states utilize groundwater protection action levels for Cr(VI), with levels ranging from 0.15 mg/kg (MI) to 100 mg/kg (LA), with a mode of 19 mg/kg (SC, TN, VT, VA, which use Region III RBCLs, and DE). Ten states utilize total chromium groundwater protective action levels, with values ranging from 2 mg/kg (NM, MS, and MT) to 350 mg/kg (OH).

Trends

As state and federal agencies seem to follow certain trends in the development of chromium soil action levels, generalizations can be made regarding the design of these criteria. First, it is important to recognize that from 1993, when we first surveyed state and federal levels, until today, the number of states utilizing soil action levels have increased from approximately 20 to 90%. Hence, the Brownfields initiatives are prompting states to develop action levels at a rapid pace. Second, the typical USEPA risk assessment model and exposure assumptions, as described in *Risk Assessment Guide to Superfund, Part A; Human Health Evaluation Manual* (USEPA, 1989a) and the previous draft SSL guidance (USEPA, 1994), are typically the basis for the state standards.* Generally, upper-bound exposure estimates (i.e., worst-case) and standard USEPA toxicity criteria are used as input parameters to ensure that the levels are protective of all land uses. Exposure factors that account for meteorological conditions and time spent away from the site are not typically included, as they are likely to vary considerably based on site-specific conditions.

One important and obvious trend, as discussed previously, is that most states and regions differentiate soil action levels by land use and chromium species. Typically, land use is separated into residential and non-residential sites, and the trivalent and hexavalent forms of chromium are evaluated separately. Most states have developed Cr(III) action levels in the range of tens of thousands of mg/kg, suggesting that Cr(III) is not considered a significant hazard.

Although a handful of states still do, most states have moved away from using Toxicity Characteristics Leaching Procedure (TCLP) to determine soil action levels for protection of groundwater. Interestingly, the USEPA SSL framework

---

* Survey was performed before the May 1996 update to SSLs. State action levels may or may not have been updated since.

recommends the Synthetic Precipitate Leaching Procedure (SPLP) leaching test rather than the TCLP for the development of SSLs protective of groundwater, yet the SPLP test alone (apart from the SSL guidance) is not identified as a soil-screening method by any state. In summary, ingestion-based action levels are about twice as frequently used as inhalation-based criteria, and inhalation criteria are about four times as frequently used as TCLP-based criteria.

Another notable trend is that states appear to be developing tiers of soil action level standards that allow for different cleanup levels with engineering controls (i.e., pavement) and institutional controls (i.e., deed and zoning restrictions) controls. Tiered approaches generally involve cleanup to a choice of (1) background, (2) media-specific statewide criteria (action levels), or (3) site-specific risk assessment. Approximately one half of the states with action levels criteria specify the use of a tiered approach.

Probably the most significant observation from this survey is that all states permit the development of remediation goals based on site-specific risk assessment. Permitting alternative cleanup standards is important for two main reasons. First, it creates a tiered set of standards for those states that do not specifically identify tiers, as the option exists to cleanup up to naturally occurring background (a position that has always been considered acceptable), state developed action levels, or site-specific risk assessment-based remediation standards. Additionally, it further secures all state-developed "actions levels" as points of departure and recognizes that site-specific alternative standards may be the most appropriate and cost-effective option for more complex sites or those with larger-scale problems.

Exceptions

A few states have adopted approaches that are exceptions to the standard risk assessment model, and as a result, have developed unusually stringent action levels. California and New Jersey do not strictly use the federal USEPA toxicity criteria as do all other states, and have developed their own for use in place of or in addition to those of the USEPA. California is the only state that considers Cr(VI) a carcinogen via both the inhalation and oral routes of exposure. All other state and federal environmental agencies consider Cr(VI) to be carcinogenic only by the inhalation route. In addition, California has developed its own cancer slope factors for both pathways. The California EPA inhalation cancer slope factor is more than an order of magnitude higher than that used by the USEPA. These conservatisms have led to an extremely conservative Cr(VI) action level of 0.2 mg/kg (the lowest of any state). In many soils, this level is below the analytical limit of detection for Cr(VI) (~1 mg/kg).

New Jersey is the only state to consider allergic contact dermatitis (ACD) as a health endpoint on which to base soil standards. In doing so, it has developed a draft Cr(VI) soil action level of 15 mg/kg (NJDEP, 1995). The New Jersey Department of Environmental Protection (NJDEP) uses USEPA toxicity criteria (i.e., reference doses and slope factors) for soil ingestion and inhalation-based standards. Additionally, it is one of two states to develop an ecological-based soil action level. New Jersey has developed a draft all soil Cr(III) standard that is protective of vegetation and phytotoxicity of 200 mg/kg, to be assessed on a site-specific basis. Minnesota has developed "ecological soil screening" levels of 0.4 mg/kg for Cr(VI) and 10 mg/kg for total chromium, which are protective of invertebrates and microbes, respectively.

Wisconsin is the only state to use an acceptable cancer risk level of 10^{-7} to derive soil action levels. This is done in an attempt to ensure that the total risk at a site with multiple chemicals in the soil is less than 10^{-6}. Every other state and USEPA region uses either 10^{-6} or 10^{-5} as acceptable cancer risk levels. Using this assumption, Wisconsin has developed the second lowest residential Cr(VI) human health-based action level of 14 mg/kg. Incidentally, California, New Jersey, and Wisconsin have the three lowest residential action levels for Cr(VI) in soil.

States that develop Cr(VI) action levels based on ingestion (i.e., based on a non-cancer endpoint) utilize a hazard index to develop their action levels. Most states use a hazard index of 1; however, both Minnesota and North Carolina use a hazard index of 0.2. Along the same lines as Wisconsin's rationale for using a 10^{-7} acceptable cancer risk level, a hazard index of 0.2 is used to account for sites with multiple non-carcinogens that produce the same critical effect (e.g., kidney damage). The hazard index of 0.2 also protects for multiple sources of exposure to the same chemical such as diet and drinking water.

New York also has a unique basis for its cleanup standards; it is the only state to develop a action level based on background or naturally occurring levels. While New York has developed risk-based standards for organic chemicals, the state's position for inorganics is that site-specific variability in the potential for inorganics to impact groundwater is so great that any concentration higher than background must be assessed on a site-by-site basis (NYDEP, 1995).

Three states, Illinois, Alaska, and Maine, have developed values for unique land uses. In addition to residential and non-residential soil standards, Illinois has also developed a Cr(VI) soil standard for construction workers. Alaska utilizes neither a "residential" nor a "non-residential" scenario, instead meteorological descriptors (i.e., arctic zone) are used to describe exposure frequency. Finally, Maine has developed draft Cr(VI) soil action criteria for a trespasser scenario in addition to the residential and non-residential scenarios.

DEVELOPMENT OF PROBABILITY DENSITY FUNCTIONS OF CR(III) AND CR(VI) HEAL THE HEALTH-BASED ACTION LEVELS (HBALs) FOR RESIDENTIAL AND NON-RESIDENTIAL SITES

As discussed previously, the majority of states and USEPA regions calculate soil action levels using methods that are either exactly the same as or modifications of the methods presented in the USEPA Soil Screening Guidance (USEPA, 1996a). To understand the variability and uncertainty associated with the state and federal action levels, probabilistic analyses of the SSL equations were performed for the soil ingestion, particulate inhalation, and protection of groundwater exposure pathways. Specifically, probability density functions (PDFs) of Cr(VI) and Cr(III) health-based action levels (HBALs) were generated by Monte Carlo statistics using PDFs of exposure parameters rather than upper-bound point estimates. While the USEPA SSLs have only been developed for residential sites, the equations were modified, using Agency guidance (USEPA, 1991b) to allow for the development of non-residential site HBALs. While the assumptions used for this analysis have been maximized to protect for non-residential sites with heavy truck traffic, these standards are also protective of commercial and industrial sites in general.

Methods

In May 1996, the USEPA finalized its methodology for calculating SSLs for three pathways: soil ingestion, inhalation, and the protection of groundwater (USEPA, 1996a). For this evaluation, the SSL equations, with only minor modifications, were used to estimate soil HBAL PDFs for Cr(VI) and Cr(III) for residential and non-residential sites via the soil ingestion and particulate inhalation pathways. In addition, the USEPA SSL equation for protection of groundwater was used to estimate HBAL PDFs for Cr(VI) and Cr(III) for that scenario.

Soil Ingestion Pathway

Because neither Cr(VI) or Cr(III) are considered oral carcinogens by any health agency, with the exception of the California EPA, a residential child scenario was evaluated for noncarcinogenic effects due to oral exposure (e.g., soil ingestion). Children have the highest soil ingestion rate to body weight ratio, and, hence, are more likely than adults to exceed the subthreshold reference dose (RfD). For the non-residential HBAL, an adult worker scenario was evaluated. For both scenarios, the following equation, identical to the corresponding SSL equation (USEPA, 1996a), was used to estimate HBAL PDFs for the soil ingestion pathway:

$$HBAL_{SI} = \frac{THQ * BW * AT}{\left(\dfrac{1}{RfD_{oral}}\right) * SIR * EF * ED * CF} \tag{1}$$

where $HBAL_{SI}$ is health-based action level for soil ingestion pathway (mg/kg); THQ is target hazard quotient (1); RfD_{oral} is USEPA chronic oral Reference Dose for Cr(III) and for Cr(VI) (mg/kg/d); SIR is soil ingestion rate (mg/d); BW is body weight (kg); AT is averaging time (days); ED is exposure duration (years); EF is exposure frequency (days/year); and CF is conversion factor (1×10^{-6} kg/mg).

The PDFs used for each of the exposure variables are presented in Table 4. The data used to develop the exposure parameter PDFs are discussed below.

Oral Toxicity Criteria. The current USEPA chronic oral RfDs for Cr(VI) and Cr(III) presented in the Integrated Risk Information System (IRIS) were used to estimate the HBALs for this pathway (USEPA, 1996c). The oral RfD for Cr(VI) of 0.005 mg/kg/d is based on a study conducted by Mackenzie *et al.* (1958) in which no effects were observed in rats administered Cr(VI) in drinking water at a concentration of 25 ppm. The oral RfD for Cr(III) of 1.0 mg/kg/d is based on a rat feeding study in which no effects were observed among rats administered Cr(III) (as insoluble chromic oxide) for 2 years at >1400 mg/kg/d (Ivankovic and Preussman, 1975; USEPA, 1996c).

Soil Ingestion Rate. Incidental soil ingestion occurs at all ages as a result of hand-to-mouth contact. It is widely believed that children approximately 2 to 5 years old are the only age group that consumes a potentially significant amount of soil (Calabrese *et al.*, 1989; van Wijnen *et al.*, 1990). Calabrese and Stanek (1991) validated soil ingestion estimates for only two of eight tracers (Zr and Ti) used in the Calabrese *et al.* study (1989). Because the confidence interval for Zr was more narrow than that for Ti, Calabrese and Stanek (1991) concluded that the child soil ingestion estimates calculated for Zr were the most accurate. Therefore, based on the Zr data, a cumulative distribution was constructed that represents daily soil ingestion for children using the median (16 mg/d), 5th percentile (–70 mg/d), 10th percentile (–35 mg/d), 90th percentile (67 mg/d), 95th percentile (110 mg/d), and maximum (1391 mg/d) (Calabrese *et al.*, 1991; Finley *et al.*, 1994). This soil ingestion rate distribution was truncated at zero and used to estimate the HBALs for the residential scenario.

There are little or no reliable quantitative data available for estimating adult soil ingestion rates. Current USEPA risk assessment guidance (USEPA, 1991b) suggests a soil ingestion rate of 50 mg/workday for adults in non-residential settings, based on the results reported in Calabrese's preliminary adult soil ingestion study (Calabrese *et al.*, 1990). However, Calabrese and Stanek (1991) have since determined that the soil ingestion rates reported in their preliminary adult study were invalid, and that the 50 mg/d value is likely to be an overestimate (Calabrese *et al.*, 1991). It is reasonable to expect that adults only eat a fraction of what children ingest (Calabrese *et al.*, 1991), probably less than 10 mg/d, as suggested by Paustenbach (1987). Accordingly, the soil ingestion rate distribution used for the child was multiplied by a factor of 0.5 and used as the soil ingestion rate PDF for adults at non-residential sites. The use of this fraction will likely overestimate the adult intake in our judgement, and the resulting PDF will be conservative.

TABLE 4
Exposure Parameter PDFs for the Soil Ingestion Pathway

Parameter	Residential		Non-residential	
	Distribution	Basis	Distribution	Basis
Body Weight (kg)	Lognormal $\mu = 14.9$ $\sigma = 4.0$	Finley et al., 1994	Lognormal $\mu = 71.0$ $\sigma = 15.9$	Finley et al., 1994
Exposure frequency (days/year)	Truncated normal $\mu = 186$ $\sigma = 58.4$	USEPA, 1989	Triangular $\mu = 245$ min $= 156$ max $= 307$	USEPA, 1989
Exposure duration (years)	Cumulative 50%ile $= 2.9$ 95%ile $= 13$	Finley et al., 1994	Cumulative 50%ile $= 3.8$ 95%ile $= 29$	Finley et al., 1994
Soil ingestion rate (mg/d)	Cumulative 50%ile $= 16$ 95%ile $= 110$	Finely et al., 1995	Cumulative (0.5 times residential) 50%ile $= 8$ 95%ile $= 55$	Best estimate

Body Weight. For this assessment, body weight data for all ages and both sexes from the second National Health and Nutrition Examination Survey (NHANES II), conducted from February 1976 through February 1980, were used to construct body weight PDFs for children (ages 0.5 to 6 years) and adults (age greater than 18 years). For the residential scenario, PDFs based on age-specific body weights that were constructed by Lloyd and Burmaster (1994) from the NHANES II data set were combined to form a lognormal PDF of body weight for a child (0.5 to 6 years) with a mean of 14.9 kg and standard deviation of 4.0 kg. For the non-residential scenario, a PDF for both adult men and women was constructed from those presented by Brainard and Burmaster (1992) using the NHANES II data. This sex-combined lognormal PDF is presented in Finley *et al.* (1994) and has a mean of 71 kg and standard deviation of 15.9 kg.

Exposure Frequency. Typically, only a fraction of soil ingested by a person is from a particular source. Consistent with USEPA guidance (USEPA, 1989a), the fraction of a day spent in contact with soil is site-specific and may be estimated based on the activity patterns of the exposed population. For a residential scenario, the exposure frequency can be considered equivalent to the fraction of the year spent at home, excluding 8-h sleeping periods as soil ingestion only occurs during waking hours. National statistics on activity patterns have been developed by the University of Michigan Institute for Social Research in terms of hours per day for adults. These data were converted from hours per day to days per year and were used to develop a truncated normal PDF with a mean of 186 d/year and a standard deviation of 58.4 d/year. As this time- activity survey does not specifically include children, the time spent at home by a child was considered to be similar to the time spent at home by his or her mother.

For the non-residential scenario, a triangular distribution based on professional judgement regarding the range of days per year an American worker is typically at the job site was constructed based on best-estimate assumptions. Assuming a full-time worker spends 5 d/week and takes 3 weeks of vacation, sick leave, and holiday time per year, a triangular PDF with a mode of 245 d is generated. The minimum of 156 d is selected for a part-time worker who works 3 d/week and has no additional vacation, sick leave, or holiday time. The maximum of 307 d is based on a full-time worker who works 6 d/week and takes only 1 week of leave per year.

Exposure Duration. For the residential scenario, the PDF for residential occupancy from birth developed by Finley *et al.* (1994), based on data collected by Johnson and Capel (1992), was used. As the distribution type for this PDF is not known, a cumulative distribution was used to generate the HBAL PDF for this scenario. This PDF has a 50th percentile of 2.9 years and a 95th percentile of 13 years. To model child and adult exposure duration, if the exposure duration was less than or equal to 6 years, all exposures were assumed to occur as a child. If the exposure duration was greater then 6 years, child exposure duration was set equal to 6 years and adult exposure was assumed to occur during the remaining years. For the non-

622

residential scenario, the PDF for occupational tenure presented in Finley *et al.* (1994), based on U.S. Bureau of Labor Statistics (1992) data, was used. This cumulative PDF has a 50th percentile of 3.8 years and a 95th percentile of 29 years. The HBALs for noncarcinogenic effects were calculated assuming exposures are not averaged over the lifetime of the individual.

Averaging Time. The HBALs for the soil ingestion pathway are based on noncarcinogenic health effects, the averaging times for both the residential and non-residential scenarios were set equal to their respective values of exposure duration (i.e., averaging time [total number of days of exposure] = exposure duration [number of years of exposure] × 365 d/year).

Particulate Inhalation Pathway

The HBAL equation used to calculate inhalation-based Cr(VI) and Cr(III) soil HBALs is a mathematical modification of the corresponding SSL equation. First, the HBAL equation was rewritten to include the inhalation rate and body weight parameters and the Cr(VI) inhalation cancer potency factor or Cr(III) inhalation RfD. While this manipulation incorporates greater uncertainty and variability into the analysis (with the addition of two exposure parameters), it allows for consistency of approach for soil ingestion and inhalation pathways. Further, these two parameters are used by the agency as point estimate parameters in the development of toxicity criteria and thus are inherent in the SSL analysis.

Second, for the non-residential HBAL, particulate emission factors were estimated assuming particulate emissions from vehicle traffic rather than wind erosion. As Cr(VI) is an inhalation carcinogen (USEPA, 1996c) and Cr(III) is not, two different HBAL equations were used. The HBALs equations for the residential and non-residential scenarios for both Cr(VI) and Cr(III) are presented below.

For Cr(VI), for the residential scenario:

$$HBAL_{PI} = \frac{Risk * AT}{CPF * IF * EF * \dfrac{1}{PEF}}$$

$$IF = \left(\frac{IR_{child} * ED_{child}}{BW_{child}} \right) + \left(\frac{IR_{adult} * ED_{adult}}{BW_{adult}} \right)$$

For Cr(VI), for the non-residential scenario:

$$HBAL_{PI} = \frac{Risk * BW * AT}{CPF * IR * EF * ED * \dfrac{1}{PEF}}$$

For both scenarios, for Cr(III):

$$HBAL_{PI} = \frac{THQ * BW * AT}{\dfrac{1}{RfD_{inhal}} * IR * ED * EF * \dfrac{1}{PEF}}$$

where $HBAL_{PI}$ is health-based action level for particulate inhalation (mg/kg); Risk is acceptable risk level (1×10^{-6}); THQ is target hazard quotient (1); RfD_{inhal} is chronic inhalation reference dose for Cr(III) (mg/kg/d); CPF is USEPA Inhalation Cancer Potency Factor for Cr(VI) $(mg/kg/d)^{-1}$; IF is age-adjusted inhalation exposure (m^3/year/kg/d); IR is inhalation rate (m^3/d); BW is body weight (kg); AT is averaging time (days); ED is exposure duration (years); EF is exposure frequency (days/year); and PEF is particulate emission factor (m^3/kg).

As Cr(VI) is carcinogenic via inhalation, these HBALs are protective of lifetime exposures as is consistent with the standard USEPA risk assessment model (USEPA, 1989a). The PDFs used for each of the exposure variables are presented in Table 5, the same PDFs that were used for body weight and exposure duration for the soil ingestion pathway were also used for the inhalation pathway. The data used to develop the exposure parameter PDFs for all other parameters are discussed below.

Inhalation Toxicity Criteria. Cr(VI) is classified as a known human inhalation carcinogen (USEPA, 1996c). Hence, soil HBALs for Cr(VI) are calculated from the USEPA Cancer Potency Factor of 41 $(mg/kg/d)^{-1}$ (USEPA, 1984). Cr(III) is not considered carcinogenic via the inhalation pathway of exposure, and non-carcinogenic toxicity via inhalation has only been observed at extremely high levels of occupational exposure (IARC, 1990; USEPA, 1990c). The USEPA has not set an RfC or acceptable exposure level for Cr(III) in ambient air. However, an inhalation RfC of 90 $\mu g/m^3$ has been estimated by Finley *et al.* (1992) based on an epidemiological study by Axelsson *et al.* (1980) of ferrochrome production workers. For this evaluation, the 90 $\mu g/m^3$ Cr(III) RfC (Finley *et al.*, 1992) was converted into dose units of mg/kg/body weight/d (0.026 mg/kg/d assuming a 70-kg adult breathes 20 m^3 per day) to calculate the Cr(III) HBAL PDFs for both scenarios.

Inhalation Rates. For the residential scenario, the PDFs for child and adult inhalation rates developed by Finley *et al.* (1994) were used. These PDFs are based on the relationship between inhalation and metabolic rates developed by Layton (1993) and are dependent on the body weight PDF. This distribution is approximately lognormal with a daily mean and standard deviation of 8.6 and 1.8 m^3/d for a child, and a daily mean and standard deviation of 15.2 and 3.3 m^3/d for an adult. Because exposures to carcinogens must be averaged over the lifetime of the individual, an "inhalation exposure" (IF) term was used for the Cr(VI) HBAL. The inhalation exposure is the age-specific inhalation rate averaged over the body weights and

TABLE 5
Exposure Parameter PDFs for the Particulate Inhalation Pathway

Parameter	Residential				Non-residential	
	Child		Adult			
	Distribution	Basis	Distribution	Basis	Distribution	Basis
Body Weight (kg)	Lognormal $\mu = 14.9$ $\sigma = 4.0$	Finley et al., 1994	Lognormal $\mu = 71.0$ $\sigma = 15.9$	Finley et al., 1994	Lognormal $\mu = 71.0$ $\sigma = 15.9$	Finley et al., 1994
Inhalation rate (m³/d)	Lognormal $\mu = 8.6$ $\sigma = 1.8$	Finley et al., 1994	Lognormal $\mu = 15.2$ $\sigma = 3.3$	Finley et al., 1994	Lognormal $\mu = 7.5$ $\sigma = 1.0$	Finley et al., 1994
Exposure frequency (days/year)	Truncated normal $\mu = 245$ $\sigma = 38.3$	USEPA, 1989	Truncated normal $\mu = 245$ $\sigma = 38.3$	USEPA, 1989	Triangular $\mu = 245$ min – 156 max = 306	USEPA, 1989
Exposure duration[a] (years)	Cumulative 50%ile = 2.9 95%ile = 13	Finley et al., 1994	Cumulative 50%ile = 2.9 95%ile = 13	Finley et al., 1994	Cumulative 50%ile = 3.8 95%ile = 29	Finley et al., 1994
Particulate emission factor (m³/kg)	1.32×10^9	Calculated value	6.79×10^8	Calculated Value	1.13×10^8	Calculated value

[a] For the residential scenario, if ED ≤ 6 years, ED_{child} = ED, and ED_{adult} = 0 years. If ED = 6 years then ED_{child} = 6 years and ED_{adult} = ED – 6 years.

625

exposure durations of a child and an adult and is units of m^3/year/kg/d. A similar factor is used in the soil ingestion SSL equation for carcinogens to calculate the age-averaged ingestion exposure (USEPA, 1996a), but is not necessary for Cr(VI) and Cr(III) SSLs, which are, as appropriate, only protective of noncarcinogenic effects.

For the non-residential scenario, a PDF for the inhalation rate of a worker for an 8-h day was developed using a similar relationship between inhalation rate and metabolic rate from Layton (1993). This PDF is approximately lognormal with a mean and standard deviation of 7.5 and 1.0 m^3/d, respectively.

Exposure Frequency. Typically, only a fraction of particulates inhaled by a person is from a particular source. Consistent with USEPA guidance (USEPA, 1989a), the fraction of a day spent in contact with a site is site-specific and may be estimated based on the activity patterns of the exposed population. For the residential scenario, the exposure frequency for the particulate inhalation pathway can be considered equivalent to the fraction of the year spent at home, including time spent sleeping. National statistics on activity patterns have been developed by the University of Michigan Institute for Social Research in terms of hours per day for adults. These data were converted into days per year and are used to develop a truncated normal PDF with a mean of 245 d/year and a standard deviation of 38.3 d/year. As for the soil ingestion pathway, as this time-activity survey does not specifically include children, the time spent at home by a child was considered to be similar to the time spent at home by his or her mother. For the non-residential scenario, the same triangular PDF used for the soil ingestion pathway was used for the inhalation pathway. These data were converted from hours per day to days per year for this analysis.

Particulate Emission Factors. To calculate soil HBALs for the particulate inhalation pathway, a description of the concentration of chemical in air in suspended soil particulates is necessary. For this evaluation, as in the SSL equation and that used by many states, a particulate emission factor (PEF), which is the inverse of the concentration of soil particulates suspended in air, was developed for the residential and non-residential scenarios. Due to the complexity of the modeling approaches used to estimate the PEFs for the two scenarios, only point estimates were used to calculate the PEFs. The PEF for the residential scenario was estimated using the particulate emission modeling approach and default parameters presented in the USEPA Soil Screening Guidance (USEPA, 1996a) for emissions due to wind erosion. The PEF for the non-residential scenario was calculated using the particulate emission modeling approach and default parameters presented in Scott *et al.* (1997) for emissions due to vehicle traffic over unpaved areas. This section presents the equations and parameters used to estimate both the residential and non-residential PEFs.

The equation for the residential PEF is the same as the one presented in the USEPA Soil Screening Guidance (1996a):

$$PEF_{residential} = \frac{Q/C * (3,600 \; sec/hr)}{0.036 * (1 - V) * \left(\dfrac{U_m}{U_t}\right)^3 * F(x)}$$

where Q/C is air dispersion factor (g/m²/s/kg/m³); V is fraction of vegetative cover; U_m is mean annual wind speed (m/s); U_t is threshold wind speed (m/s); and F(x) is an aerodynamic function of U_m/U_t (unitless).

The same parameters used by the USEPA in the Soil Screening Guidance were used to calculate the residential PEF, and are presented in Table 6. The calculated residential PEF based on these default parameters is 1.32×10^9 m³/kg.

The PEF for the non-residential scenario was calculated using the AP–42 equation for vehicle traffic over unpaved roads (USEPA, 1995d) and the dispersion factor for a 2 acre site as presented in the Soil Screening Guidance (USEPA, 1996a). Soil suspension from wind erosion was not incorporated in this analysis because suspension from wind erosion is insignificant compared with that from truck traffic on unpaved surfaces. The generic vehicle traffic parameters for a non-residential site with an area of 10,000 m² (~2 acres) presented in Scott *et al.* (1997) were used to calculate emissions from truck traffic on unpaved sites. The equation for the non-residential site PEF is

$$PEF = \frac{Q/C * 86,000 \; sec/day * A_{site}}{1.7 * k * \left(\dfrac{s}{12}\right) * \left(\dfrac{S}{48}\right) * \left(\dfrac{W}{2.7}\right)^{0.7} * \left(\dfrac{w}{4}\right)^{0.5} * \left(\dfrac{365 - p}{365}\right) * TC * D}$$

where Q/C is air dispersion factor (g/m²/s/kg/m³); A_{site} is area of the generic site (m²); k is particle size multiplier; s is silt content (%); S is mean vehicle speed (km/h); W is mean vehicle weight (Mg); w is mean number of wheels; p is number of days with greater than 0.01 in of precipitation; TC is number of vehicles traveling across site per day (vehicles/day); and D is average distance traveled per day (km).

With the exception of the air dispersion factor and silt content, the same parameters for the generic non-residential site presented in Scott *et al.* (1997) were

TABLE 6
Parameters Used To Calculate the
Residential Particulate Emission Factor (PEF)

Parameter	Value	Ref.
Dispersion Factor (g/m²/s/kg/m³)	90.80	USEPA, 1996
Fraction of vegetation cover	0.5	USEPA, 1996
Mean wind speed (m/s)	4.69	USEPA, 1996
Threshold wind speed (m/s)	11.32	USEPA, 1996
F(x) (unitless)	0.194	Calculated

used to calculate the non-residential PEF and are presented in Table 7. The air-dispersion factor presented by the USEPA in the Soil Screening Guidance for a 10,000 m^2 site was used. The average silt content for a non-residential site, presented by Cowherd in USEPA (1989b), of 8% was also used. The calculated non-residential PEF based on these default parameters is 1.13×10^8 m^3/kg. It should be noted that the parameters used to describe the non-residential site PEF may vary dramatically based on site-specific conditions, significantly impacting the results.

Protection of Groundwater Pathway

The method used to calculate the soil HBALs for the protection of groundwater is identical to the USEPA SSL equation for the protection of groundwater (USEPA, 1996a). This equation calculates the soil concentration for a chemical that produces a soil porewater concentration equal to or less than a risk-based or an acceptable groundwater concentration. This equation is based on the commonly used and well-known linear equilibrium soil/water partition equation that describes the ability of a chemical to sorb to soil. The equation for the soil HBAL protective of groundwater is (USEPA, 1996a):

$$ HBAL_{PGW} = MCL * \left(K_D + \frac{\theta}{\rho_b} \right) * DAF $$

where HBAL$_{PGW}$ is health-based action level for the protection of groundwater (mg/kg); MCL is maximum contaminant level of Cr(VI) and Cr(III) in groundwater (0.1 mg/l for both); K_D is soil-water partition coefficient (l/kg); θ_w is water-filled soil porosity (l/l); ρ_b is soil bulk density (kg/l); and DAF is dilution/attenuation factor.

TABLE 7
Parameters used to Calculate the
Non-Residential Particulate Emission Factor (PEF)

Parameter	Value	Ref.
Dispersion factor (g/m^2/s/kg/m^3)	70.6	USEPA, 1994
Particle size multiplier	1.0	Scott et al., 1997
Silt content (%)	8	USEPA, 1989
Mean vehicle speed (km/h)	24	Scott et al., 1997
Mean vehicle weight (Mg)	15	Scott et al., 1997
Number of wheels	6	Scott et al., 1997
Days with >0.01 in of precipitation	115	World Almanac, 1989
Traffic count (vehicles/day)	10	Scott et al., 1997
Distance per trip (km)	0.1	Scott et al., 1997
Area of site (m^2)	10,000	Scott et al., 1997

The PDFs and point estimates used for each of the parameters are presented in Table 8. The point estimate value for soil bulk density of 1.5 kg/l from the USEPA Soil Screening Guidance was used. Distributions were used for all other parameters, and the data on which they are based are discussed below.

Water-Filled Porosity. The PDF for soil porosity for silty loam soils presented by Carsel and Parrish (1988) was used to generate the PDF for water-filled porosity assuming that two thirds of the pore space was taken up by water at any given time. This PDF is lognormal with a mean of 0.3 l/l and a standard deviation of 0.05 l/l and is based on the soil database for the U.S. compiled by Carsel and Parrish (1988).

Soil-Water Partition Coefficient. The PDFs for soil-water partition coefficient, or K_D, for Cr(III) and Cr(VI) are lognormal based on soil-water partition data from several studies and sites summarized by Baes and Sharp (1983) and presented by Dragun (1988). K_D values for Cr(VI) and Cr(III) demonstrate a high degree of variance with soil pH; however, the 222 samples of soil included in this summary provide data only for a "typical range" of pH for agricultural soils. In these soils, pH was found to be normally distributed with a mean of 6.7, and 95th percentiles of the pH values were between 4.7 and 8.7. Further extremes of soil pH will result in even greater variance in K_D values. For the pH range provided, the K_D values for Cr(III) and Cr(VI) had lognormal distributions with means of 4530 and 410 l/kg, respectively. The K_D PDFs represent a diverse mixture of soils, study conditions,

TABLE 8
Parameter PDFs for the Protection of Groundwater Pathway

Parameter	Distribution	Ref.
Water-filled soil porosity (l/l)	Lognormal $\mu = 0.30$ $\sigma = 0.05$	Carsel and Parrish, 1988
Soil bulk density (kg/l)	Point estimate 1.5	USEPA, 1996
Soil-water partition coefficient (KD) for Cr(III) (l/kg)	Lognormal $\mu = 4,530$ $\sigma = 8,000$	Baes and Sharp, 1983
Soil-water partition coefficient (KD) for Cr(VI) (l/kg)	Lognormal $\mu = 410$ $\sigma = 3,140$	Baes and Sharp, 1983
Dilution/attenuation factor (unitless)	Cumulative 50th = 14.5 95th = 864	USEPA, 1996
Maximum contaminant level (MCL) for Cr(III) and Cr(VI) (mg/l)	Point Estimate 0.1	USEPA, 1996

and extraction procedures. While the K_D data from these studies for Cr(III) and Cr(VI) may or may not be comparable to all sites, they represent the most current and best database for use in developing a PDF for this parameter.

Dilution/Attenuation Factor. As chemicals move through soil and groundwater, they are subjected to a number of physical, chemical, and biological processes that generally reduce their concentrations at a receptor point. This reduction in concentration can be described in a straightforward manner using a dilution/attenuation factor (DAF). The DAF is the ratio of soil leachate concentration to groundwater concentration at the receptor point. This factor will always be greater than one (no dilution) and in many cases will be quite large (>1000). The cumulative distribution of estimated DAFs for 300 hazardous waste sites located throughout the U.S. were used to generate a PDF for this parameter (USEPA, 1996a). This PDF has a 50th percentile of 14.5 and a 95th percentile of 964 mg/kg.

Cr(VI) and Cr(III) HBAL PDFs

The cumulative distribution functions (CDFs) for the HBAL PDFs for the residential and non-residential scenarios for both Cr(III) and Cr(VI), with 5th percentiles less than 1,000,000 mg/kg, are presented in Figures 1 through 7. The 5th and 50th percentiles of the HBAL distributions are presented in Table 9.

Soil Ingestion Pathway

For the residential scenario, the 5th percentile of the HBAL distribution for Cr(III) is 178,000 mg/kg, while the 50th percentile is greater than 1,000,000 mg/kg (exceeds unity) (Figure 1). For the non-residential scenario, both the 5th and 50th percentiles of the Cr(III) HBAL distribution exceeded unity. The 5th and 50th percentile of the residential soil HBAL PDF for Cr(VI) are 890 and 9410 mg/kg, respectively (Figure 3). For the non-residential scenario, the 5th and 50th percentiles of the soil HBAL PDF are 7630 and 654,000 mg/kg, respectively (Figure 1). For this pathway, the residential scenario produces lower HBALs than the non-residential scenario, indicating that residents are a more sensitive subpopulation than workers for soil ingestion exposures. This is because children have a higher soil ingestion to body weight ratio than adults and are more frequently exposed at their residence.

Particulate Inhalation Pathway

All of the HBAL PDFs for Cr(III) for this pathway were greater than unity. The 5th and 50th percentiles for the residential Cr(VI) HBAL PDF are 670 and 2120

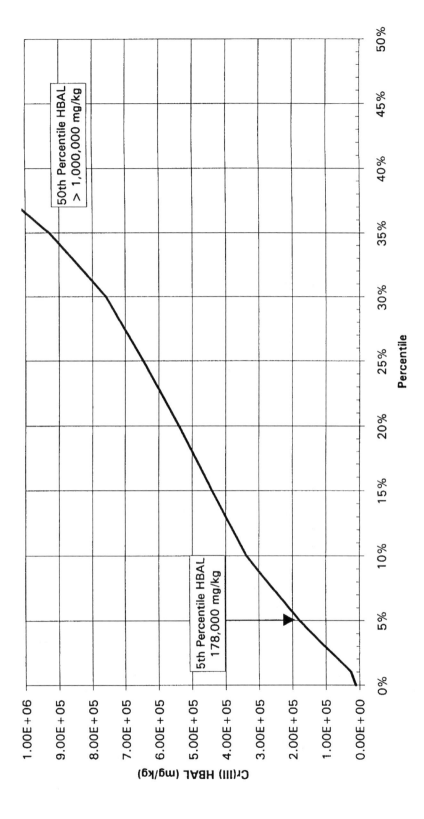

FIGURE 1

Cumulative distribution function (CDF) of Cr(III) HBALs for the residential scenario soil ingestion pathway.

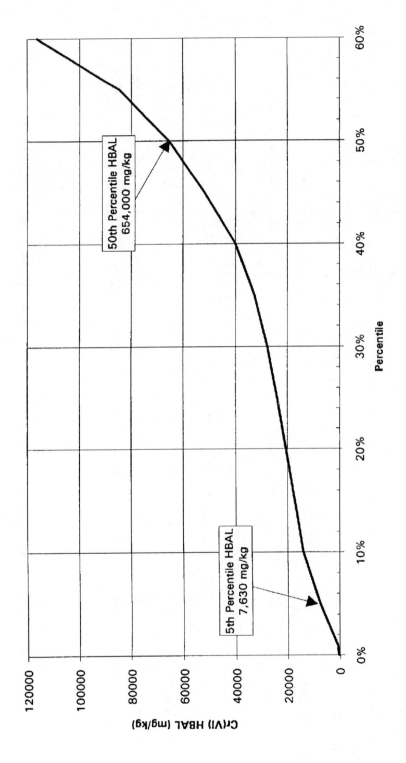

FIGURE 2

Cumulative distribution function (CDF) of the Cr(VI) HBALs for the non-residential scenario soil ingestion pathway.

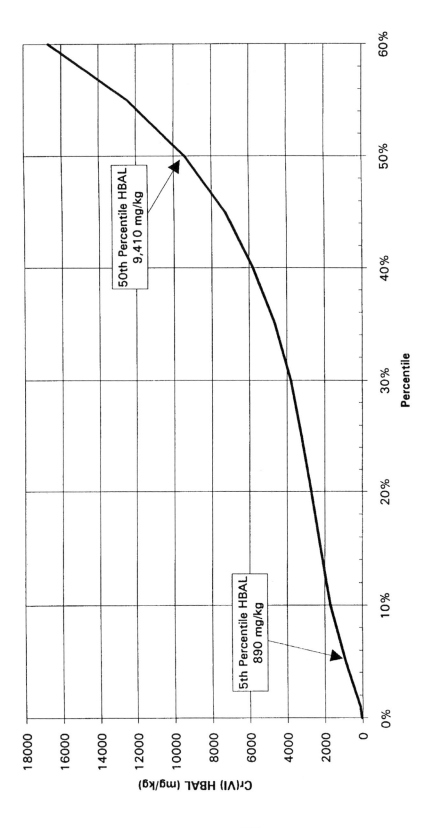

FIGURE 3

Cumulative distribution function (CDF) of the Cr(VI) HBALs for the residential scenario soil ingestion pathway.

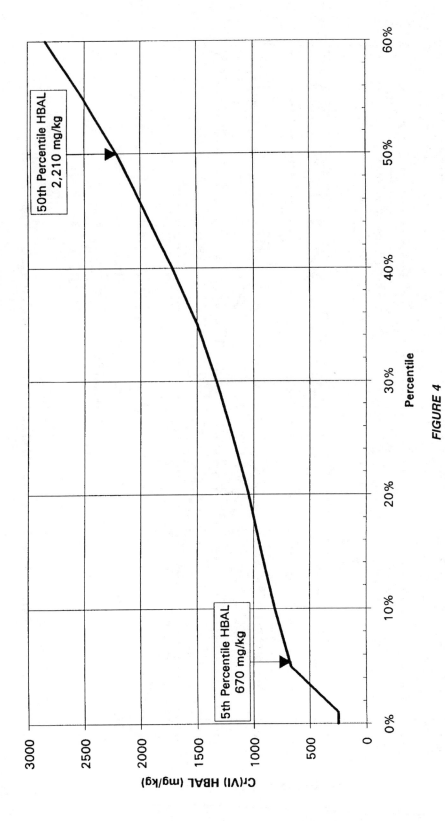

FIGURE 4

Cumulative distribution function (CDF) of the Cr(VI) HBALs for the residential scenario particulate inhalation pathway.

634

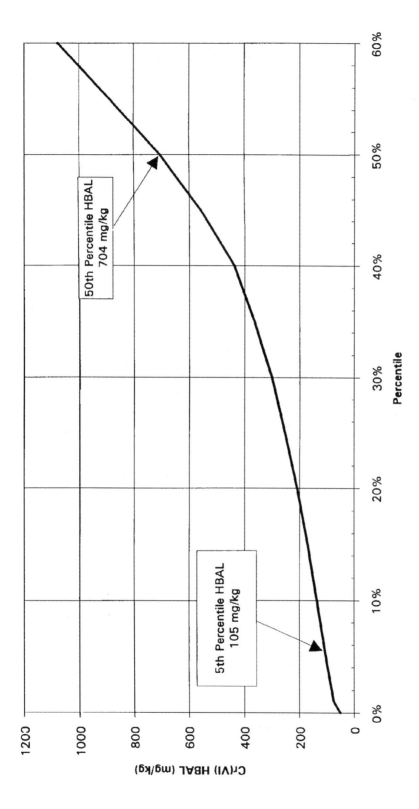

FIGURE 5

Cumulative distribution functions (CDF) of the Cr(VI) HBALs for the non-residential scenario particulate inhalation pathway.

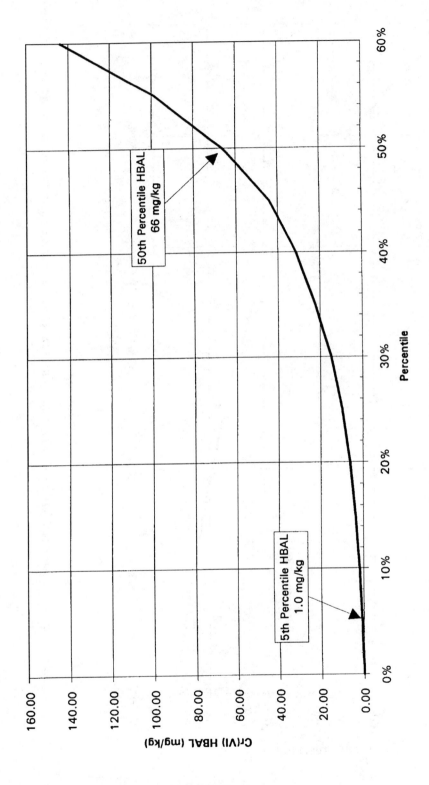

FIGURE 6

Cumulative distribution function (CDF) of Cr(VI) HBALs for the protection of groundwater.

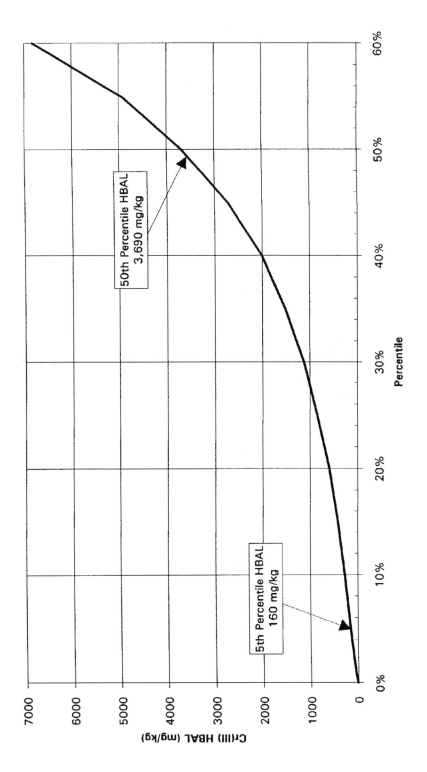

FIGURE 7

Cumulative distribution function (CDF) of Cr(III) HBALs for the protection of groundwater.

TABLE 9

Summary of PDFs of HBALs for all Pathways (mg/kg)

	Residential		Non-Residential		Protection of groundwater
Percentile	Soil ingestion	Particulate inhalation	Soil ingestion	Particulate inhalation	
			Cr(VI)		
5th	892	670	7,630	105	1.0
50th	9,410	2,210	654,000	704	66
			Cr(III)		
5th	178,000	>1,000,000	>1,000,000	>1,000,000	160
50th	>1,000,000	>1,000,000	>1,000,000	>1,000,000	3,690

mg/kg, respectively (Figure 4). For the non-residential scenario, the 5th and 50th percentiles for the Cr(VI) HBAL PDF are 105 and 704 mg/kg, respectively (Figure 5). For this pathway, the HBALs for the non-residential scenario are lower than those for the residential scenario. This is opposite of the relationship between residential and non-residential HBALs for the soil ingestion pathway and is a result of the higher concentration of airborne particulates generated by vehicle traffic at non-residential sites vs. the concentration generated by wind erosion alone at residential sites.

Protection of Groundwater

The 5th, 50th, and 95th percentiles of the Cr(III) PDF of HBALs for the protection of groundwater are 160, 3690, and 499,000 mg/kg, respectively (Figure 7); for Cr(VI), the 5th, 50th, and 95th percentiles were 1.0, 66, and 19,710 mg/kg, respectively (Figure 6). The variability in HBALs for this pathway is much greater than that for the soil ingestion and particulate inhalation pathways. The primary reason for the greater variability is the high variability and uncertainty associated with the PDFs for the soil-water partition coefficient (K_D) and the dilution/attenuation factor (DAF). The variability for these parameters is high because the site-specificity of both is very high.

DISCUSSION

The 5th percentile of the HBAL PDF is theoretically protective of 95% of possible exposures; thus, it likely provides reasonable yet conservative HBALs for screening concentrations of Cr(III) and Cr(VI) in soil. The lowest HBALs for each chromium species for each land use are 178,000 and 670 mg/kg, respectively, for Cr(III) and Cr(VI) at residential sites, and the lowest HBAL for Cr(VI) at non-residential sites is 105 mg/kg. No HBAL for Cr(III) at non-residential sites is necessary as the calculated values exceed unity. These HBALs are protective of direct contact human exposures; however, the potential hazards associated with chemical migration from soil to groundwater and potential impacts to ecological receptors must still be made on a site-by-site basis. It should be noted, however, that these potential hazards are unlikely to be significant for either Cr(III) or Cr(VI) at most sites as discussed below.

The suggested HBALs are thought to be reasonable, because they were generated by probabilistic analysis (Monte Carlo statistics) that avoids the problem of compounded conservatism associated with selecting the worst-case point estimate for each exposure parameter (Finley *et al.*, 1994). Further, the HBALs are protective of human contact exposure scenarios because they are the lower bound (5th percentile) of the PDF, are calculated with conservative USEPA toxicity criteria, and do not consider additional site-specific parameters, which in all cases will further limit exposure. These include extreme meteorological conditions (i.e., snow, ice, rain, etc.), physical barriers (i.e., pavement, buildings, and landscaping), and limits on the bioaccessibility of chromium from the soil matrix (i.e., release of chromium from soil in biological fluids to allow for systemic absorption).

Comparison of 5th Percentile HBALs with State and Federal Action Levels

The Cr(VI) non-residential HBAL of 105 mg/kg is lower than most of the 29 Cr(VI) non-residential state action levels (Plate 1*). The HBAL is lower than those developed by most states because many of the state standards do not consider the inhalation of particulates pathway, and only half of those evaluate the potential for traffic on unpaved surfaces. Traffic or other physical disturbances are a potentially significant source of soil suspension, in addition to wind erosion. While the HBAL of 105 mg Cr(VI)/kg is probably too conservative for use as a cleanup level for most non-residential sites, because of the multiple conservatisms in the analysis, it provides a "point of departure" from which a site-specific risk assessment could be performed. Because the HBAL is based on soil suspension to ambient air and long-term (chronic) inhalation exposures averaged over a lifetime, the HBAL is comparable to the mean concentration of Cr(VI) in soil or another measure of central tendency.

Interestingly, the non-residential Cr(VI) HBAL of 105 mg/kg is lower than both the USEPA inhalation SSL of 270 mg/kg (for residential scenarios) and the Region III transfer from soil to air RBCL of 140 mg/kg, as well as the residential HBAL of 670 mg/kg, calculated herein. This is because they all are based on wind erosion alone. Thus, in the case of potential Cr(VI) exposures, assuming the residential land use scenario is not always the most health-protective assumption. This occurs because Cr(VI) is only carcinogenic via inhalation, and the potential inhalation dose is larger at an industrial/commercial site. The USEPA should consider placing an additional limitation on the Cr(VI) SSL, as it may not be protective for sites with significant mechanical disturbance of surface soil (i.e., truck traffic, tilling or plowing, construction activities). This maybe the case for other inorganics (i.e., arsenic, cadmium) that are recognized to exhibit greater toxicity via the inhalation route when compared with the oral route of exposure.

The Cr(VI) residential inhalation-based HBAL calculated here (670 mg/kg) is notably higher than the modes (390 mg/kg) of the state action levels for Cr(VI) for residential sites. Thirty-eight state action levels are lower than 670 mg/kg, and only

---

* Plate 1 appears following page 641.

640

four states have levels that are higher (Plate 1). The HBAL calculated here for soil ingestion (892 mg/kg) is more than two times higher than the mode of the state levels (390 mg/kg), reflecting the increased conservatism of the approaches used by the states when compared with that used here.

A comparison of the Cr(VI) inhalation-based SSL (270 mg/kg) to the 5th percentile of the residential inhalation HBAL PDF (670 mg/kg) clearly demonstrates that the incorporation of the probabalistic analysis, and use of the body weight and inhalation rate PDFs with the cancer slope factor, rather than simply the unit risk factor, more than doubles the calculated level. This is significant considering that the same equation and particulate emission factor were used. Hence, the incorporation of a probabalistic analysis into the USEPA SSL framework increases the calculated concentrations by avoiding compounded conservatism. Because of the increased level of effort and expertise associated with the execution of a probabilistic analysis, its incorporation into the SSL framework could limit availability and usefulness of SSLs to the public. However, the financial benefits associated with improved risk management decisions, and the reduction of remedial efforts, may be substantial when considered nationally and should be weighed against these drawbacks. Only one state, Ohio, has met the challenges associated with incorporating probabilistic analysis in their action levels. However, some of the action levels developed by Ohio EPA, based on a probabilistic analysis, are among the lowest, indicating that the incorporation of probabilistic analysis into the process does not necessarily increase action levels.

The HBAL for Cr(III) at residential sites of 178,000 mg/kg is higher than most (20 of 23) state action levels (Plate 2*). Nonetheless, most Cr(III) action levels for residential sites (including the mode of the state and federal action levels of 78,000 mg/kg) are very high. Levels of Cr(III) this high are rarely encountered in soil, perhaps only in certain slags from specialty steel manufacturing that can contain high concentrations of chromium, or concentrated waste steams from chromium-using industries such as chrome plating or leather tanning. Thus, it is unlikely that an HBAL for Cr(III) is actually necessary, even for residential sites. Site-specific concerns for wildlife, including vegetation, and the potential for impacts to groundwater should still be considered for Cr(III) in soil, particularly in acidic environments in which Cr(III) may be more readily solubilized (Saleh *et al.,* 1989).

As this analysis suggests, insoluble forms of Cr(III), for which the USEPA RfD is set (USEPA, 1996c), are clearly not a significant human health concern in soil at almost all sites. Further, insoluble forms of Cr(III), such as chromic oxide and chromic sulfate, are unlikely to pose a hazard to groundwater or potential ecological receptors. In both cases, insolubility restricts the ability of these Cr(III) compounds to be transported in the environment and affect biological organisms.

---

*   Plate 2 appears following page 641.

Other Pathways of Exposure

Pathways of exposure that are not typically evaluated for the development of soil action levels for chromium include ingestion of home-grown vegetables for a residential scenario; ingestion of milk, beef, and fish, and dermal contact with soil. None of these pathways are generally a significant concern for Cr(III) or Cr(VI).

Ingestion of Milk, Beef, Fish, or Vegetables.   Ingestion of mother's milk, cow milk, beef, or fish is unlikely to pose a potential human health concern due to chromium accumulation from soil because neither species concentrates in the muscle or fat of humans or animals (Langard, 1982; Klaassen *et al.*, 1986). Additionally, neither species concentrates in edible plant tissues (fruits and vegetables) (Kabata-Pendias and Pendias, 1984; Levi *et al.*, 1973), and, therefore, ingestion of vegetables or fruit grown in chromium containing soils is not a human health concern. In fact, in an attempt to ease the problem of chromium nutritional deficiency in the U.S., researchers have attempted to enrich agricultural soils with chromium but have not been successful (Desmet *et al.*, 1975; Cary *et al.*, 1977 a,b; Lahouti and Peterson, 1979; Ramachandran *et al.*, 1980; Cary and Kubota, 1990).

Dermal Contact with Soil.   Neither Cr(VI) or Cr(III) are absorbed well across the skin. Systemic absorption of Cr(III) across intact skin is negligible (Mali *et al.*, 1963; Polak *et al.*, 1983) and the absorption rate of freely soluble Cr(VI) is low (Wahlberg and Skog, 1965; Wahlberg, 1970; Baranowska-Dutkiewicz, 1981). Further, because only chromium in solution can be dermally absorbed, the issue of bioaccessibility, or extraction of chromium from soil by human sweat on the skin, will substantially limit the potential dose associated with dermal contact with chromium in soil. The fraction of Cr(VI) that may be released from COPR by real human sweat has been estimated to be <0.1% (Horowitz and Finley, 1993). Waimen *et al.* (1994) measured 12% extraction of Cr(VI) from COPR with artificial sweat. The substantial difference between the two measurements is probably due to the organic component of real sweat that is absent from artificial sweat (Horowitz and Finley, 1993). Nonetheless, it may be assumed that dermal bioaccessibility of Cr(VI) from soil will be quite limited.

The New Jersey Department of Environmental Protection has determined a draft standard of 10 ppm (mg/L) Cr(VI) in solution in the environment (i.e., a puddle) to protect against allergic contact dermatitis (ACD) (NJDEP, 1995). Considering that this concentration is 2 orders of magnitude greater than the MCL (0.1 ppm) and 3 orders of magnitude higher than the freshwater chronic ambient water quality criteria for Cr(VI) (0.011 ppm), the usefulness of an environmental water standard as high as 10 ppm Cr(VI) for any health endpoint is uncertain. It is likely that other health end points, such as the inhalation cancer risk or the protection of ground-water, for soluble forms of Cr(VI), will dictate soil cleanup standards much lower than concentrations that could generate a standing water concentrations as high as

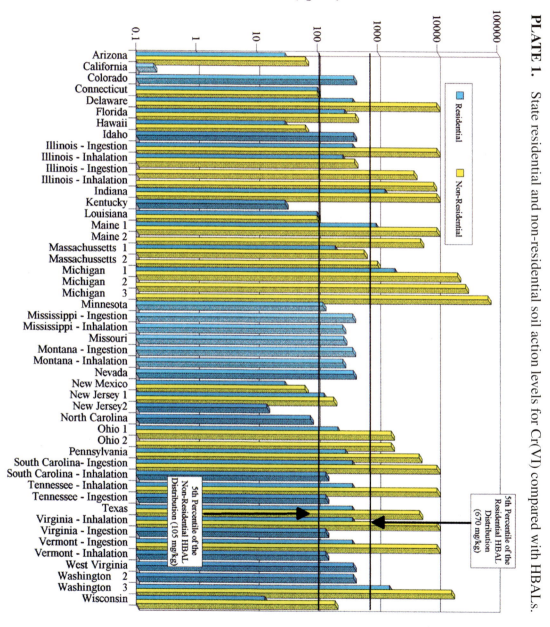

PLATE 1. State residential and non-residential soil action levels for Cr(VI) compared with HBALs.

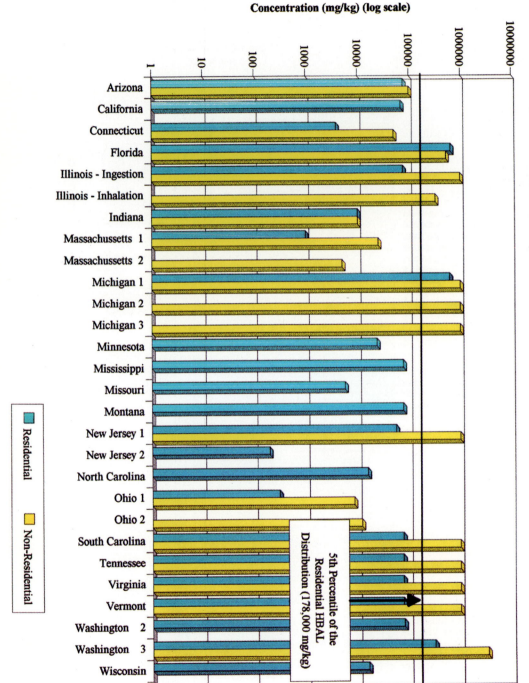

PLATE 2. State residential and non-residential soil action levels for Cr(III) compared with HBAL.

10 ppm Cr(VI). Further, it should be noted that the scientific defensibility of this threshold value for dermal exposure (10 ppm) has been questioned primarily because it is based on historical patch-testing studies, where dose was reported in terms of concentration of Cr(VI) in the testing solution or patch test rather than mass per area (Paustenbach *et al.*, 1992; Nethercott *et al.*, 1994; Crump, 1995; Sielken, 1995; Felter and Dourson, 1997).

Nethercott *et al.* (1994) performed a human patch testing study to determine the 10% response threshold for elicitation of ACD among previously sensitized individuals in terms of mass of Cr(VI) per surface area of skin. Using the USEPA risk assessment model for evaluating dermal exposure to soil (USEPA, 1992), these researchers determined that 450 mg/kg Cr(VI) in soil would be necessary to elicit ACD in 10% of sensitized individuals. This value is lower than the residential HBAL (670 mg/kg); however, it is based on the conservative assumption that 100% of the Cr(VI) is extracted from soil in sweat or solution on the skin. Chromium extraction studies have shown the bioaccessability of chromium to sweat to be less that 0.1% from COPR (Horowitz and Finley, 1993). Therefore, ACD should not be used as a basis for soil action levels for Cr(VI); however, it may be a consideration in site-specific environmental or occupational health risk assessment.

Site-Specific Considerations

It is important to emphasize that the HBALs calculated here, as well as those calculated by the states and regions (with the exception of New Jersey's draft 200 mg/kg Cr(III) level protective of phytotoxicity and the Minnesota ecological screening levels of 0.4 mg/kg Cr(VI) protective of invertebrates and 10 mg/kg total Cr protective of microbes) are not specifically protective of ecological receptors. Further, these HBALs are for soil and are not comparable to sediments unless direct human contact with the sediments is the only pathway for exposure.

Ecological Concerns. The Ontario Ministry of Ecology and Environment (OMEE) has determined severe and low-effect levels of 110 mg/kg and 26 mg/kg, respectively, for total chromium in sediment (OMEE, 1993). Long *et al.* (1995) have determined low and medium effect range levels of 81 and 370 mg/kg, respectively, total chromium, and the State of Washington (Bureau of National Affairs, Inc. 1995) has developed a marine sediment quality criterion of 260 mg/kg for total chromium. While there is a high degree of uncertainty associated with these criteria, a straightforward comparison of these sediment criteria to the soil HBALs suggests that sites with the potential for chromium sediment contamination, but soil concentrations below appropriate action levels, should be investigated for potential ecological impacts. However, it should be noted that much higher levels of total chromium in sediments (as high as 1760 mg/kg) have been shown not to pose an ecological hazard for either aquatic toxicity or bioconcentration in the Hackensack River (Hall and Pulliam, 1993). Generally, one would expect that the bioavailability of chromium in sediments would be very low as the high organic content and neutral pH of sediments converts Cr(VI), or soluble Cr(III), into insoluble chromic(III)oxides (Saleh et al., 1989).

Groundwater. The HBAL distribution calculated for protection of groundwater for both Cr(VI) and Cr(III) varied widely (3 to 5 orders of magnitude between the 5th and 95th percentiles). This is because the propensity of either species to leach to groundwater is highly dependent on site-specific conditions, including the solubility of the chemical form of chromium present, the soil chemistry that dictates leaching potential (pH, organic carbon content, porosity), and the hydrogeological conditions that dictate DAF values. For these reasons, an action level for protection of groundwater is not recommended. Rather, leaching to groundwater must be evaluated on a site-by-site basis. The approaches offered by the Illinois EPA (IEPA, 1996) and the USEPA SSL framework (USEPA, 1996a) for incorporating a limited amount of site-specific information into the calculation of action levels in tiers (layers of analysis of increasing complexity) provide useful tools for site-specific assessments.

SUMMARY AND CONCLUSIONS

In general, state and federal agencies have used innovative approaches, based on the current USEPA risk assessment paradigm, for developing non-site-specific health-based action levels for Cr(III) and Cr(VI). Based on a probabilistic analysis of the USEPA framework for determining SSLs, HBALs for Cr(VI) of 105 mg/kg, and 670 mg/kg for non-residential and residential sites, respectively, were determined. An HBAL for Cr(III) for residential sites of 178,000 mg/kg was also determined; no such corresponding HBAL for Cr(III) at non-residential sites was deemed necessary. These HBALs are somewhat higher, although generally consistent in approach with those developed by the states and federal USEPA regions. It is important to incorporate soil suspension modeling and the inhalation pathway of exposure when setting action levels for Cr(VI). Although this pathway contributes only a small portion to the total dose, the carcinogenic potential of Cr(VI) via inhalation causes standards for this pathway to be the lowest (most health protective). Useful HBALs for Cr(VI) and Cr(III) for protection of groundwater cannot be calculated as the estimates vary too widely based on site-specific conditions. The USEPA Soil Screening Guidance, which allows for the incorporation of site-specific information, appears to be the most practical as well as scientifically defensible approach for calculating action levels.

Just one benefit of the Brownfields initiatives is the development of soil action levels to facilitate environmental cleanup and reclamation of otherwise unused resources. The action levels are already decreasing perceived or real burdens associated with environmental regulation for both the regulators and regulated community. Ultimately, the benefits of the Brownfields initiatives will be realized by the community through a cleaner environment, expanded tax base, new jobs, and an overall revitalization of, sometimes almost deserted, industrial areas.

REFERENCES

Adams, R. M. 1990. Job descriptions with their irritants and allergens. *Occupational Skin Disease,* 2nd ed. pp. 666–668. Philadelphia, W. B. Saunders Company.

Agency for Toxic Substances and Disease Registry (ATSDR). 1994. Toxicological Profile for Chromium. U.S. Department of Health and Human Services. Prepared by Clement Associates under contract No. 205-88-0608.

Axelsson, G., Rylander, R. and Schmidt, A. 1980. Mortality and incidence of tumors among ferrochromium workers. *Br. J. Ind. Med.* **37,** 121–127.

Baes, C. F., III and Sharp, R. D. 1983. A proposal for estimation of soil leaching and leaching constants for use in assessment models. *J. Environ. Qual.* **12(1),** 17–28.

Baranowska-Dutkiewcz, B. 1981. Absorption of hexavalent chromium by skin in man. *Arch Toxicol* **47,** 47–50.

Brainard, J. and Burmaster, D. E. 1992. Bivariate distributions for height and weight of men and women in the United States. *Risk Anal.* **12(2),** 267–275.

Bureau of National Affairs, Inc. 1991. Washington Sediment Management Standards. Washington, D.C.

Burmaster, D. E. and Lehr, J. H. 1991. It's time to make risk assessment a science. Editorial. *Ground Water Monit. Rev.* Summer.

Calabrese, E. J., Barnes, R., Stanek, E. J., Pastides, H., Gilbert, C. E., Veneman, P., Wang, X., Lasztity, A., and Kostecki, P. T. 1989. How much soil do young children ingest: an epidemiologic study. *Reg. Toxicol. Pharm.* **10,** 123–137.

Calabrese, E. J., Stanek, E. J., Gilbert, C. E. and Barnes, R. M. 1990. Preliminary adult soil ingestion estimates: results of a pilot study. *Reg. Tox. Pharm.* **12,** 88–95.

Calabrese, E. J., E. J. Stanek and C. E. Gilbert. 1991. A preliminary decision framework for deriving soil ingestion rates. In: *Hydrocarbon Contaminated Soils and Groundwater: Analysis, Fate, Environmental and Public Health Effects,* Remediation Volume (Kostecki, P. T., Calabrese, E. J., and Bell, Eds.).

Calabrese, E. J. and Stanek, E. J. 1991. A guide to interpreting soil ingestion studies. *Regul. Toxicol. Pharm.* **13,** 278–292.

Carsel, R. F. and Parrish, R. S. 1988. Developing joint probability distributions of soil water retention characteristics. *Water Res. Res.* **245,** 755–769.

Cary, E. E., Allaway, W. H., and Olson, O. E. 1977a. Control of chromium concentrations in food plants. I. Absorption and translocation of chromium by plants. *J. Agric. Food Chem.* **25(2),** 300–304.

Cary, E. E., Allaway, W. H., and Olson, O. E. 1977b. Control of chromium concentrations in food plants. II. Chemistry of chromium in soils and its availability to plants. *Agric. Food Chem.* **25(2),** 305–309.

Cary, E. E. and Kubota, J. 1990. Chromium concentration in plants: effects of soil chromium concentration and tissue contamination by soil. *J. Agric. Food Chem.* **38(1),** 108–114.

Cowherd, D., Muleske, G. E., Englehart, P. J., and Gillette, D. A. 1985. Rapid Assessment of Exposure to Particulate Emissions from Surface Contamination Sites. U.S. Environmental Protection Agency, Office of Health and Environmental Assessment. EPA/600–8–85/002.

Crump, D. I. K. 1995. Comments on the Benchmark Methodology Used by New Jersey Department of Environmental Protection (NJDEP) in the Document *Derivation of a Risk-Based Elicitation Response Concentration for Hexavalent Chromium Cr(VI) Based on Allergic Contact Dermatitis.*

Desmet, G., De Ruyter, A., and Ringoet, A. 1975. Absorption and metabolism of $(CrO_4)^{2-}$ by isolated chloroplasts. *Phytochemistry* **14**, 2585–2588.

Dragun, J. 1988. The fate of hazardous materials in soil what every geologist and hydrologist should know. II. *Hazardous Materials Control.* **13**, 41–65.

Felter, S.P. and Dourson, M. L. 1997. Hexavalent chromium-contaminated soil: options for risk assessment and risk management. *Regul. Toxicol. Pharmacol.* **25**, 43–59.

Finkel, A. M. 1990. Controlling Uncertainty in Risk Management. A Guide for Decision Makers. Centers for Risk Management, Resources for the Future. Washington, D.C.

Finley, B. L. Proctor, D. M., Scott, P. K., Harrington, N., Paustenbach, D., and Price, P. 1994. Recommended distributions for exposure factors frequently used in health risk assessment. *Risk Analysis* **14(4)**, 533–553.

Finley, B. L., Proctor, D. M., and Paustenbach, D. J. 1992. An alternative to the USEPA's proposed inhalation reference concentrations for hexavalent and trivalent chromium. *Regul. Toxicol. Pharmacol.* **16**, 161–176.

Hall, W. S. and Pulliam, G. W. 1995. An assessment of metals in an estuarien wetlands ecosystem. *Arch. Environ Contam Toxicol.* **29**, 164–173.

Horowitz, S. B. and Finley, B. L. 1993. Using human sweat to extract chromium from chromite ore processing residue: applications to setting health-based cleanup levels. *J. Toxicol. Environ. Health* **40**, 585–599.

Illinois Environmental Protection Agency (IEPA). 1996. Tiered Approach to Cleanup Objectives Guidance Document, January, 1, 1996.

International Agency for Research on Cancer. IARC. (1990). IARC Monographs on the evaluation of carcinogenic risks to humans. *Chromium, Nickel and Welding.* Vol. 49. IARC Working Group on the Evaluation of Carcinogenic Risks to Humans, Lyon, France, June 5–13, 1989, World Health Organization.

Ivankovic, S. and Preussmann, R. 1975. Absence of toxic and carcinogenic effects after administration of high doses of chromic oxide pigment in subacute and long-term feeding experiments in rats. *Food Cosmet. Toxicol.* **13**, 347–351.

Johnson, J. and Capel, J. 1992. Monte Carlo Approach to Simulating Residential Occupancy Periods and its Application to the General U.S. Population USEPA Office of Air Quality Planning and Standards, EPA–450/3–92-011, Washington, D.C., 1992.

Kabata-Pendias, A. and Pendias, H. 1984. *Trace Elements in Soils and Plants.* Boca Raton, FL, CRC Press.

Klaassen, C. D., Amdur, M. O., et al. 1986. *Casarett and Doull's Toxicology: The Basic Science of Poisons.* New York, MacMillan Publishing Company.

Lahouti, M. and Peterson, P. J. 1979. Chromium accumulation and distribution in crop plants. *J. Sci. Food Agric.* **30**, 136–142.

Langard, S. 1982. Adsorption, transport, and excretion of chromium in man and animals. In: *Biological and Environmental Aspects of Chromium, Topics in Environmental Health* pp. 149–169.

Layton, D. W. 1993. Metabolically consistent breathing rates for use in dose assessment. *Health Physics* **641**, 23–26.

Levi, E., Dalschaert, X., and Wilmer, J. B. M. 1973. Retention and absorption of foiliar applied Cr. *Plant Soil* **38**, 683–686.

Lloyd, K. J. and Burmaster, D. E. 1994. Lognormal distributions of body weight for female and male children in the United States. *J. Expos. Anal. Environ. Epid.* submitted for publication.

Long, E. R., MacDonald, D. D., Smith, S. L., and Calder, F. D. 1995. Incidence of adverse biological effects within ranges of chemical concentrations in marine and estuarine sediments. *Environ. Manage.* **19**(1), 81–97.

MacKenzie, R. D., Byerrum, R. U., Decker, C. F., Hoppert, C. A., and Langham, R. F. 1958. Chronic toxicity studies. II. Hexavalent and trivalent chromium administered in drinking water to rats. *Am. Med. Assoc. Arch. Ind. Health* **18**, 232–234.

Mali, J. W. H., Van Kooten, W. J., and Van Neer, F. C. J. 1963. Some aspects of the behavior of chromium compounds in the skin. *J. Invest. Dermatol.* **41**(I), 111–122.

Mertz, W. 1969. Chromium occurrence and function in biological systems. *Physiol. Rev.* **49**(2), 163–239.

Nethercott, J., Paustenbach, D., Adams, R., Fowler, J, Marks, J., Morton, C., Taylor, J., Horowitz, S., and Finley, B. 1994. A study of chromium induced allergic contact dermatitis with 54 volunteers: implications for environmental risk assessment. *Occup. Environ. Med.* **51**(6), 371–380.

New Jersey Department of Environmental Protection (NJDEP). 1995. Draft Site Remediation Program. September 1995.

New York Department of Environmental Protection (NYDEP). 1995. Letter from Michael J. O'Toole, Jr. Division of Hazardous Waste Management to Deborah Proctor and Kurt Fehling of McLaren/Hart, Pittsburgh, PA.

Ontario Ministry of Environment and Energy (OMEE). 1993. Development of the Ontario Provincial Sediment Quality Guidelines for Arsenic, Cadmium, Chromium, Copper, Lead, Manganese, Mercury, Nickel, and Zinc. Prepared by R. Jaagumagi, Water Resources Branch.

Paustenbach, D. J. 1987. Assessing the Potential Environmental and Human Health Risks of Contaminated Soil. *Comments Toxicol.* **13–4**, 185–226.

Paustenbach, D. J., Meyer, D. M., Sheehan, P. J., and Lau, V. 1991. An assessment and quantitative uncertainty analysis of the health risks to workers exposed to chromium contaminated soils. *Toxic. Ind. Health* **7**, 159–196.

Paustenbach, D. J., Sheehan, P. J., Paull, J. M., Wisser, L. M., and Finley, B. L. 1992. Review of the allergic contact dermatitis hazard posed by chromium-contaminated soil: identifying a "safe" concentration. *J. Toxic. Environ. Health.* **37**, 177–207

Polak, L. 1983. Immunology of chromium. In: *Chromium: Metabolism and Toxicity.* pp. 69–72. (Burrows, D. Ed.) Boca Raton, CRC Press.

Ramachandran, V., D'Souza, T. J., and Mistry, K. B. 1980. Uptake and transport of chromium in plants. *J. Nucl. Agric. Biol.* **9**, 126.

Saleh, F. Y., Parkerton, T. F., Lewis, R. V., Huang, J. H., and Dickson, K. L. 1989. Kinetics of chromium transformations in the environment. *Sci. Total Environ.* **86**, 25–41.

Scott, P. K., Sung, H. M., Finley, B. L., Schulze, R. H., and Turner, B. D. 1997. Identification of an accurate soil suspension/dispersion modeling method for use in estimating health based soil cleanup level of hexavalent chromium in chromite-ore processing residues. *J. Air Waste Manage. Assoc.* **47**: 753–765.

Sielken, R. L. 1995. Comments to the New Jersey Department of Environmental Protection Proposed Hexavalent Chromium Soil Standard Based on Elicitation of Allergic Contact Dermatitis in Sensitized Individuals. October 8, 1995.

United States Department of Labor. 1992. Employee Tenure and Occupational Mobility in the Early 1990s. Bureau of Labor Statistics, Washington, DC, USDL 92–386.

U.S. Environmental Protection Agency (USEPA). 1984. Health Assessment Document for Chromium Final Report. U.S. Environmental Protection Agency, Environmental Criteria and Assessment Office, Research Triangle Park, NC. EPA–600/8–83-014F. August 1984.

U.S. Environmental Protection Agency (USEPA). 1989a. Risk Assessment Guidance for Superfund, Volume 1. Human Health Evaluation Manual Part A, interim final. Office of Emergency and Remedial Response, Washington, D.C. EPA/540/1–89/002.

U.S. Environmental Protection Agency (USEPA). 1989b. Exposure Factors Handbook. Office of Health and Environmental Assessment, Washington, D.C.

U.S. Environmental Protection Agency (USEPA). 1990a. National Oil and Hazardous Substances Pollution Contingency Plan. U.S. Environmental Protection Agency. 55FR 8666. March 8, 1990.

U.S. Environmental Protection Agency (USEPA). 1990b. Test Methods for Evaluating Solid Wastes, Physical/Chemical Methods. SW–846, Third Edition, Final Update, Office of Solid Waste and Emergency Response, Washington, D.C.

U.S. Environmental Protection Agency (USEPA). 1990c. Noncarcinogenic Effects of Chromium Update to Health Assessment Document.

U.S. Environmental Protection Agency (USEPA). 1991a. National Primary Drinking Water Regulations; Final Rule. January 30, 1991. *Fed. Reg*: 3525–3597.

U.S. Environmental Protection Agency (USEPA). 1991b. Human Health Evaluation Manual, Supplemental Guidance: "Standard Default Exposure Factors." Directive 9285.6-03. Interim Final. Office of Emergency and Remedial Response, Toxics Integration Branch, Washington, D.C.

U.S. Environmental Protection Agency (USEPA). 1992. Guidance for Exposure Assessment. *Fed. Reg.* **59(104)**, 22888–22936.

U.S. Environmental Protection Agency (USEPA). (1994). *Draft Soil Screening Guidance*. Office of Solid Waste and Emergency Response. Office of Emergency and Remedial Response; Hazardous Site Control Division. December, EPA

U.S. Environmental Protection Agency (USEPA). 1995a. Risk-Based Concentration Table. Memorandum from Roy L. Smith, Senior Toxicologist, Region III.

U.S. Environmental Protection Agency (USEPA). 1995b. Region IX Preliminary Remediation Goals (PRGs) Second Half 1995. San Francisco, CA.

U.S. Environmental Protection Agency (USEPA). 1995c. Proposed Rule. July 25, 1995. *Fed. Reg*: 37974.

U.S. Environmental Protection Agency (USEPA). 1995d. Compilation of Air Pollutant Emission Factors, Fifth Edition. Office of Air Quality Planning and Standards.

U.S. Environmental Protection Agency (USEPA). 1996a. Soil Screening Guidance: Technical Background Document. Office of Solid Waste and Emergency Response. Office of Emergency and Remedial Response; Hazardous Site Control Division. May, EPA/540/R–95/128.

U. S. Environmental Protection Agency (USEPA). 1996b. Brownfields Action Agenda. Washington, DC.

U. S. Environmental Protection Agency (USEPA). 1996c. Integrated Risk Information Service (IRIS). U.S. Environmental Protection Agency USEPA, Cincinnati, OH.

U.S. Department of Health, Education and Welfare. 1962. Public Health Service Publication No. 956.

Van Wijnen, J. H., Clausing, P., and Brunekreef, B. 1990. Estimated soil ingestion by children. *Environ. Res.* **51**, 147–162.

Vitale, R., Mussoline, G., Petura, J., and James, B. 1993. A Method Evaluation Study of an Alkaline Digestion Modified Method 3060 Followed by Colorimetic Determination Method 7196A for the Analysis for Hexavalent Chromium in Solid Matrices. Environmental Standards, Inc., Applied Environmental Management, Inc., University of Maryland, Laboratory of Soil Chemistry.

Wahlberg, J. E. 1970. Percutaneous absorption of trivalent and hexavalent chromium ^{51}Cr through excised human and guinea pig skin. *Dermatologica* **141**, 288–296.

Wahlberg, J. E. and Skog, E. 1965. Percutaneous absorption of trivalent and hexavalent chromium. *Arch. Dermatol.* **92**, 315–318.

Wainman, T. and Hazen, R., et al. 1994. The extractability of Cr(VI) from contaminated soil in synthetic sweat. *J. Expos. Anal. Environ. Epid.* **4**, 171–181.

Journal of Soil Contamination, 6(6):649–705 (1997)

Using Applied Research to Reduce Uncertainty in Health Risk Assessment: Five Case Studies Involving Human Exposure to Chromium in Soil and Groundwater

Brent L. Finley[1] and Dennis J. Paustenbach[2]

[1]*Chemrisk®, a Division of McLaren/Hart, Inc., 29225 Chagrin Boulevard; Cleveland, Ohio 44122;* [2]*McLaren/Hart Environmental Engineering, 1135 Atlantic Avenue, Alameda, Ca 94501*

In this article, five case studies are presented that involve original research conducted in order to better understand the potential health risks associated with human exposure to Cr(VI) in soils and groundwater. Each study was designed to address a specific data gap, and all of these studies involved the use of human volunteers and/or the study of human biological fluids. The results of this research can be summarized as follows: (1) soil concentrations of approximately 1240 ppm Cr(VI) or less do not elicit allergic contact dermatitis (ACD) in a vast majority of the general population (>99.9%), and soil concentrations much higher than this value are also health protective if the Cr(VI) is not readily bioavailable; (2) exposure to soil concentrations up to 400 ppm total chromium is unlikely to influence urinary chromium levels; (3) the human gastrointestinal tract can reduce ingested Cr(VI) to Cr(III) at concentrations up to 10 mg Cr(VI)/l; and (4) at water concentrations of up to approximately 22 mg Cr(VI)/l, dermal penetration of Cr(VI) is negligible even under extreme exposure conditions. Based on these results, it appears that: (1) ACD is not an appropriate health endpoint for setting health-based soil standards, (2) in many cases, urinary biomonitoring studies are unlikely to be useful in assessing Cr(VI)-related exposures, (3) the USEPA's MCL of 0.10 mg Cr(VI)/l contains a large margin of safety, and (4) systemic uptake of Cr(VI) following dermal contact with water or soil does not occur to a degree that warrants quantitative evaluation in a health risk assessment. The results obtained from carefully designed human volunteer studies generally do not contain the inherent uncertainties associated with extrapolation from animal or in vitro studies. If the work can clearly be performed at no risk to the participants, then consideration should be given to using human subjects in the design and conduct of risk assessment research.

KEY WORDS: *human exposure, chromium. research, risk assessment.*

1058-8337/97/$.50

$\mathcal{A}$ quantitative health risk assessment is an integrated analysis that draws on numerous scientific disciplines: toxicology, epidemiology, low-dose extrapolation, statistics, pharmacokinetics, chemical fate and transport modeling, Monte-Carlo analysis, exposure assessment, and others (Paustenbach, 1995). Because the practice of health risk assessment is still relatively new, much of the necessary information is "borrowed" from these various disciplines and, as a result, the critical estimates and assumptions that form the basis of the analysis have usually been developed for some purpose other than for use in health risk assessment. For example, very few animal exposure studies have ever been conducted for the express purpose of developing toxicity criteria (e.g., reference doses and cancer slope factors) to evaluate human noncancer and cancer risks associated with exposure to environmental chemicals. Most toxicity criteria are based on studies intended for some other objective, and therefore they are often lacking in some respect (i.e., insufficient dose duration, inadequate number of doses, failure to identify a no-effect level, etc.). Yet, it is widely acknowledged that toxicity criteria introduce a large degree of uncertainty in health risk assessment (e.g., Paustenbach, 1989, 1995). The same shortcomings exist to some degree in all of the information sources that are used to create a health risk assessment. Accordingly, the accuracy of the risk assessment process has justifiably been questioned in recent years.

In response to the uncertainties associated with the lack of "risk-assessment specific" data, there has been a growing trend toward conducting focused research specifically to diminish or eliminate uncertainty in some of the key parameters in a risk assessment. The accuracy of the exposure assessment process, in particular, has increased dramatically in the last 10 years primarily as a result of research that has improved our understanding of the frequency, duration, and magnitude with which humans come in contact with environmental media. Exposure estimates that were once based solely on intuition or educated guesses have largely been supplanted by quantitative measurements obtained in laboratory or field studies designed specifically to better define a particular exposure parameter. For example, a 1984 risk assessment for Times Beach (Kimbrough *et al.*, 1984) relied primarily on professional judgement to estimate soil ingestion rates for children (10,000 mg/d), which, at the time, was considered to be the most significant exposure pathway. Since then, numerous investigators have used fecal tracer studies to quantitatively measure soil ingestion rates in young children (e.g., Davis *et al.*, 1990; Calabrese *et al.*, 1989; Calabrese and Stanek, 1991; Calabrese and Stanek, 1995), and the uncertainty associated with the preliminary estimates has largely been diminished, as has the estimated rate itself (upper-bound value of 200 mg/d). Likewise, in the last 10 years, over 15 studies have measured soil adherence on skin (e.g., Roels *et al.*, 1980; Duggan *et al.*, 1985; Kissel *et al.*, 1996), and further refinement in these estimates would probably only slightly increase the accuracy of assessing dermal contact rates with soil (Finley *et al.*, 1994). Indeed, for many of the exposure parameters that govern the primary and secondary exposure pathways (inhalation rates, ingestion rates for water, garden-grown vegetables, and mother's milk, dermal contact with surface water, etc.), sufficient data exist to allow the risk assessor to develop a reasonable estimate for most site-specific situations. In short, the general level of refinement that has been reached in exposure assessment suggests that at least some of our risk assessment resources should be directed toward research in other areas of significant uncertainty.

One critical source of uncertainty in most risk assessments is our understanding of the degree to which environmental contaminants cross biological barriers and distribute systemically within tissues. Such data simply do not exist for many common environmental contaminants. For the most part, risk assessors typically rely on surrogate data or general assumptions that are based on little quantitative information (e.g., the oft-used assumption that metals in soils are 1% dermally bioavailable). Hence, although risk assessors have become relatively proficient at estimating how much media we come in contact with, we are not as adept at estimating the degree to which the chemical moves out of the media, comes in contact with and moves across biological barriers (skin, lung, and gastrointestinal tract), and distributes into the various tissue compartments where the "effects" ultimately occur. As the practice of health risk assessment continues to move (albeit slowly) toward assessing potential health effects as a function of chemical concentration at the target organ, rather than the current "applied/absorbed dose" metric, it will become even more critical to better understand how chemicals move from environmental media into tissues.

Hexavalent chromium [Cr(VI)] provides a unique opportunity to explore how applied research can reduce uncertainty in this area. Hexavalent chromium is a common environmental contaminant associated with numerous industrial processes. It is currently classified by the U.S. Environmental Protection Agency (USEPA) as an inhalation carcinogen, and dermal exposure to soluble forms of Cr(VI) in the workplace can elicit allergic contact dermatitis (ACD). Although the presence of Cr(VI) in soils and groundwater often "drives" a risk assessment as the primary chemical of interest, very little is known regarding its bioavailability from environmental matrices, its behavior after contact with biological barriers, or how it distributes into human tissues following systemic absorption. The fact that Cr(VI) is known to reduce to nontoxic Cr(III) after contact with reducing media, such as biological tissues, further complicates the analysis.

In this article, five case studies are presented that involve original research conducted in order to better understand the possible health risks associated with human exposure to Cr(VI) in soils and groundwater. The first three studies were designed to specifically address risk assessment uncertainties associated with exposure to Cr(VI) in soil in Hudson County, NJ. This particular problem, which involves several million tons of chromite-ore processing residue (COPR), has posed several challenging technical issues with which risk assessors have grappled over the last 5 years. Well over 30 papers have been published recently concerning the proper conduct of COPR-related health risk assessments and more are forthcoming. The latter two case studies evaluate general exposure to contaminated tapwater, which is a potentially significant primary exposure pathway at numerous sites in the U.S. All of the five studies were designed to address a specific data gap, and each involved the use of human subjects and/or the study of human biological fluids. Each investigation is preceded by a brief overview of the background information available at the time, the uncertainty that was identified, the manner in which the research program was conducted, and how the results were interpreted to address the data gap. Each study then concludes with a discussion of how the findings should influence current Cr(VI) health risk assessments.

CASE STUDY #1: Identifying a Minimum Elicitation Threshold For Cr(VI): Implications for Setting Soil Standards Protective of Individuals with Allergic Contact Dermatitis

BACKGROUND

Several million tons of chromite ore were processed in Hudson County, NJ, between 1897 and 1971 (Environmental Science and Engineering [ESE], 1989). This procedure involved roasting a chromium-rich ore in a kiln with soda ash or lime and then leaching it with water. The remaining "residue" material contained from 2 to 5% total chromium, primarily in two forms: hexavalent chromium [Cr(VI)] and trivalent chromium [Cr(III)] (ESE, 1989). Between 1905 and 1960, approximately 3 million tons of this chromite-ore processing residue (COPR) were used to fill in low-lying areas of Hudson County to provide base material for industrial building pads and parking lots, for tank dikes, and to serve as backfill following demolition activities. Total chromium [Cr(VI) and Cr(III)] concentrations in COPR measured at these sites ranged from 5 to approximately 70,000 ppm, and Cr(VI) concentrations ranged from <0.5 to approximately 7000 ppm (ESE, 1989). In recent years, there has been a significant amount of discussion among regulatory authorities and health risk assessors regarding the nature and extent of the health hazards posed by the COPR and the need for health-based cleanup standards for Cr(III) and Cr(VI) (New Jersey Department of Environmental Protection [NJDEP], 1990; Bagdon, 1989; Paustenbach et al., 1991a,b, 1992; Sheehan et al., 1991).

Cr(VI) is one of the most common dermal sensitizers in occupational settings and accounts for about 5% of clinically reported cases of ACD in the U.S. (Marks et al., 1992). Cr(VI)-related ACD, which has been reported in chromium plating workers (Burrows, 1983; 1990), lithographers (Levin et al., 1959), diesel repair shop workers (Winston et al., 1951), and leather workers (Morris, 1958), is known as a type IV, delayed, or cell-mediated allergic reaction (Marks and DeLeo, 1992; Burrows, 1983; Adams, 1990). The localized biological response of ACD is

652

similar to a "poison oak" hypersensitive reaction and elicits the standard symptoms of erythema, edema, and small vesicles (Ackerman, 1978; Adams, 1990; Marks and DeLeo, 1992). Type IV allergic dermatitis reactions are most often not life threatening, and their effects are generally limited to the skin. Epidermal contact with high concentrations of Cr(VI) can also produce irritant contact dermatitis (ICD), a nonimmunological response (Rook *et al.,* 1986; Jackson and Goldmen, 1990; Lammintausta and Maibach, 1990).

Several patch testing studies were conducted in the 1950s and 1960s to determine the "minimum elicitation threshold" (MET) of Cr(VI) in sensitized persons (Pirilä, 1954; Anderson, 1960; Geiser *et al.,* 1960; Zelger, 1964; Burrows *et al.,* 1965; Skog and Wahlberg, 1969). In those studies, people known to be Cr(VI) sensitive were tested with patches containing serial dilutions of Cr(VI) [usually as potassium dichromate ($K_2Cr_2O_7$)] in petroleum jelly, water, and/or acid glycine. The NJDEP evaluated these historical studies in an effort to establish a soil concentration that would prevent elicitation of ACD in Cr(VI)-sensitized individuals who might come in contact with the Cr(VI)-impacted soils. They suggested that an elicitation threshold of approximately 10 ppm Cr(VI) could be derived from the dose-response data (Bagdon and Hazen, 1991; NJDEP, 1992a). Based on this apparent "threshold," NJDEP proposed a 10 ppm Cr(VI) soil standard for the several million tons of COPR in New Jersey.

However, as described in Paustenbach *et al.* (1992), much of the data on which NJDEP relied were collected before the improved and standardized diagnostic criteria developed by the North American Contact Dermatitis Group (NACDG) and the International Contact Dermatitis Research Group (ICDRG) (NACDG, 1984), and it is highly likely that some irritant reactions were scored as allergic responses in these earlier studies. Also, these reports often failed to disclose information regarding the diagnostic criteria to determine allergy, duration of patch application, and the analytical methods used to validate the chromium concentration and valency state. Further, it is known that the patch preparation methods were inconsistent and that interpatch variability of the amount of Cr(VI) applied could be as high as an order of magnitude (Fischer and Maibach, 1985). However, the influence of all of these shortcomings pale in comparison to the one significant deficiency that renders these data meaningless for the purposes of setting ACD-based soil standards: the patch concentrations were always reported in terms of mass of Cr(VI) per mass of patch (i.e., "ppm") and not in terms of mass per unit area (e.g., μg-Cr(VI)/cm²-skin). As described in detail in Horowitz and Finley (1994), area-based elicitation thresholds (μg-Cr(VI)/cm²-skin) are required in order to properly estimate a soil concentration of Cr(VI) that will not elicit ACD following dermal contact with soil. Prior to 1994, these data simply did not exist.

Case Study #1 describes the results of a patch test study designed to determine the area-based MET (μg-Cr(VI)/cm²-skin) of solubilized Cr(VI) that will elicit ACD in Cr(VI)-sensitized subjects. Under the direction of several members of the NACDG, a group of 54 participants known to be sensitized to Cr(VI), were patch tested with serial dilutions of Cr(VI). To reduce the variability inherent in earlier

653

patch preparation methods, "TRUE-Test" patches were specifically manufactured for use in this study. Additional detail concerning this study may be found in Nethercott *et al.* (1994).

MATERIALS AND METHODS

In general, as the water solubility of a Cr(VI) salt increases, its ability to penetrate the skin barrier and elicit ACD increases (Gammelgaard *et al.*, 1992). Due to its high degree of water solubility, $K_2Cr_2O_7$ is one of the most penetrating and therefore most potentially reactive Cr(VI) species (Paustenbach *et al.*, 1992). Potassium dichromate is currently used by members of the NACDG and ICDRG as the standard Cr(VI) patch test agent for diagnostic purposes (Rietschel *et al.*, 1989; Storrs *et al.*, 1989; ICDRG, 1994). Accordingly, for the purposes of this study, $K_2Cr_2O_7$ was chosen as the Cr(VI) test compound.

TRUE-Test (thin layer rapid use epicutaneous test) gel matrix patches were manufactured by Kabi Pharmacia Research Center AS, Inc. (Hillerod, Denmark). Patches were prepared by mixing $K_2Cr_2O_7$ (purity 98.5 to 101.5%) with a wet hydroxy propyl cellulose gel to the specified concentrations. These gels have little or no sensitizing potential (Fischer and Maibach, 1985; 1989; Storrs *et al.*, 1989). The $K_2Cr_2O_7$ was mixed to a specified mass of Cr(VI) per unit area, printed on a sheet of polyester, and dried to a thin film. These coated water impermeable sheets were then cut into square patches of 0.81 cm², mounted on a piece of adhesive, non-allergenic tape, and packaged in an air-tight and light-impermeable envelope. The $K_2Cr_2O_7$ test patches were also sealed with desiccant paper to prevent adsorption of moisture.

The TRUE-Test patches are specifically designed to hydrate by perspiration when taped to the skin under occlusion. The dried film is hydrated into a gel thickness of 50 to 70 μm, from which the allergen migrates to the skin. The hydrated gel, occluded by adhesive backing and plastic, ensures maximal contact with the skin, thus enabling high allergen bioavailability (Rietschel, 1989). With this approach, the allergen is evenly distributed over the test area, and the quantitative dose of allergen challenge is accurately controlled (Fischer and Maibach, 1984; 1985; Wright, 1991). This provides a significant advantage over other current techniques, such as the Finn chamber, in which the mass of allergen loaded on to the skin may vary by up to an order of magnitude from patch to patch (Fischer and Maibach, 1985).

STUDY DESIGN

Patch Concentrations

The Cr(VI) patch concentrations used in this study were chosen based on a best estimate of a range that would provide a maximal (100%) response at the highest concentration and a minimal response (<10%) at the lowest concentration. A review of the literature suggested that an approximate 250-fold range of Cr(VI) concentrations from 0.018 to 4.4 μg-Cr(VI)/cm^2 would yield a complete dose-response curve. The following Cr(VI) concentrations were used: 0.018, 0.088, 0.18, 0.88, and 4.4 μg-Cr(VI)/cm^2.

Patch Testing Strategy

The patch testing study was designed to occur in three rounds. In the first round all subjects were tested with a diagnostic Cr(VI) patch (4.4 μg-Cr(VI)/cm^2) to confirm that all volunteers in the study were allergic to Cr(VI). Those who were confirmed as sensitized in round one were then tested in round two with the two lowest Cr(VI) concentrations (0.018 and 0.088 μg-Cr(VI)/cm^2). If the subjects responded to both of the Cr(VI) patches or only to the higher concentration (0.088 μg-Cr(VI)/cm^2), a threshold was considered to have been identified and the subjects did not complete round three. If no response occurred at either of the two low Cr(VI) concentrations, then the subjects were tested in round three with the two higher Cr(VI)-concentrations (0.18 and 0.88 μg-Cr(VI)/cm^2). For each subject, the lowest Cr(VI) concentration at which a positive response occurred was considered to be the MET for that subject.

The rationale for testing with low Cr(VI) concentrations initially, followed by higher concentrations if necessary, was to minimize the incidence of "false positives" and "excited skin syndrome" that can occur when multiple patches are applied to the subject's skin in a single dosing (Maibach, 1981, 1987; Bruynzeel *et al.*, 1983). Indeed, many of the earlier reported positive responses to low Cr(VI) concentrations almost certainly were false-positives resulting from multiple testing (Maibach, 1981, 1987; Bruynzeel *et al.*, 1983; Nethercott, 1982). All test patches were supplied to the physicians under code to double blind the study. The purpose of blinding was to achieve unbiased readings of the test results by the study physicians. Only one physician evaluated each patient. The physicians received approval from their respective human use committees as appropriate.

Population Size

Before starting the study, an analysis was performed to estimate the total number of subjects required to achieve acceptable statistical power. The goal was to identify a lower threshold at which no more than 10% of the Cr(VI)-sensitive subjects would respond with 90% confidence. A figure of 10% was used because a number of dermatologists indicated that it would be very difficult to accurately identify through testing a 1 to 5% value and because 10% is probably very close to the threshold dose for all people (including the most highly sensitized). Estimating the number of subjects needed for the study required an assumption about the number of responders at a particular test concentration. Based on results of previous reports (Bagdon, 1989), it was concluded that about 50 to 80 subjects would be required to meet these criteria.

It was anticipated that a number of the subjects may have actually had irritant, atopic, or "excited skin syndrome" reactions rather than a true ACD response during previous testing and that only about 80% of the initial subjects would, on round one testing, show a positive response to Cr(VI). Of these, it was estimated that at least 75% of sensitized people would consent to participate in the proposed studies and complete the required testing rounds. The drop out rate for these subjects was not known but was not expected to exceed 25%. In summary, to have a study population of no less than 50 subjects, it was determined that about 100 Cr(VI)-sensitized volunteers were necessary.

Participating Physicians and Study Population

Participating Physicians. Six practicing dermatologists conducted the clinical aspects of the study (Dr. Robert Adams, Dr. Joseph Fowler, Dr. James Marks, Dr. Charles Morton, Dr. James Nethercott, and Dr. James Taylor). They also participated in the study design.

Study Population. More than 6000 patient files from various dermatologists (who specialized in patch testing) were examined before 100 possible volunteers were identified. Eventually, a group of 113 potential subjects was found by the participating physicians, of which 102 eventually took part in the study (11 subjects dropped out due to personal reasons). All were believed to be Cr(VI)-sensitized based primarily on previous clinical patch tests performed by these physicians. As presented in Table 1, this initial study population consisted of 78 men (76%) and 24 women (24%). All were over 18 years of age. People taking immunosuppressive or steroidal medications and pregnant women were excluded from the study. Subjects with eczema at the scheduled time of testing were not

TABLE 1
Description of Cr(VI) Sensitized
Volunteers who Participated in the Study

Sex	Number originally Patch tested	Number who qualified and participated	Range of ages (y)	Average age (y (SD))
Men	78	39	24–74	45.6 (12.6)
Women	24	15	25–59	39.6 (10.3)

tested until the dermatitis subsided, and it was requested that topical steroids not be used for 2 weeks before testing. All volunteers provided their doctors with written consent to participate in the study.

Patient Questionnaire. Before initial testing, each patient filled out a medical and occupational history questionnaire. Each questionnaire was screened by a qualified person to ensure proper completion. The medical history discussed in the form included incidence, type, and duration of past and present dermatitis and other known allergies, including asthma, and any other skin problems or sensitivities. The questionnaire also asked for history of jobs held and corresponding duration, as well as any known or potential exposure to Cr in the workplace. Use of over-the-counter medications and vitamins was also recorded.

Information on allergic and atopic dermatitis was available for all of the 102 subjects. Present or past atopic dermatitis was present in 15% of those who completed the study (8 people). Twenty-two percent of the volunteers worked in the construction industry or a related field, and a significant number of participants had no known previous occupational exposure to Cr(VI) (Table 2).

Patch Testing Procedure

In each round of testing, all patches were applied to the upper sides of the back at 7 cm apart and fixed with Scanpor or paper tape for total occlusion, with each patient serving as his or her own control. The patches remained in place for 48 h, at which time they were removed and readings were taken then and at 96 h. Test sites with positive reactions after 96 h were photographed. For each patch in each round, the physicians recorded one of the following responses:

1 = Weak (no vesicular) reaction: erythema, infiltration, papules (+)

2 = Strong (edematous or vesicular) reaction (++)

3 = Extreme (spreading, bullous, ulcerative) reaction (+++)

4 = Doubtful reaction, macular erythema only (?)

5 = Irritant morphology

6 = Negative reaction (–)

7 = Not tested

Patch Test Results

Round One. Of the 102 people who were initially selected for the study, only 54 responded positively (+, ++, or +++) to the diagnostic Cr(VI) patch (4.4 µg-

TABLE 2
Summary of Occupations of the 54 Cr(VI)
Sensitized Volunteers Who Participated in the Study

Occupation	No. of people
Accounting	2
Art Dealer	1
Brick Mason	3
Carpenter	2
Carpet Layer	1
Cement Mason/Finisher	4
Truck Driver (Concrete)	1
Core Driller	1
Electrician	1
Engineer	2
Gardener	1
Horse Trainer	1
Insulation Installer	1
Lab Technician	2
Mechanic	1
Medical Assistant	1
Photoengraver	3
Plumber (Retired)	1
Print Estimator	1
Production Manager	2
Publisher	1
Real Estate Associate	1
Retired	5
Saw Cutter	1
Secretary	5
Self Employed	1
Student	1
Teacher/Educator	3
Tire Changer	1
Transit Operator	1
Warehouse Worker	1
Unemployed	1

Cr(VI) cm^2). This response rate (53%) was far less than the projected estimate of 80% but still yielded enough people (54) to provide statistically significant results. The 54 subjects who proceeded to round two of the study consisted of 39 men (72%) and 15 women (28%). The men ranged in age from 24 to 74 (mean 45.6) years; the women from 25 to 59 (mean 39.6) years (Table 1).

Round Two. In round two, four of 54 subjects (7%) responded positively to 0.088 µg-Cr(VI)/cm^2 but not to 0.018 µg-Cr(VI)/cm^2. Accordingly, a MET of

0.088 µg-Cr(VI)/cm^2 was recorded for these four people, and they were not required to complete round three. Only one of 54 (2%) responded positively to both 0.018 µg-Cr(VI)/cm^2 and 0.088 µg-Cr(VI)/cm^2, a MET of 0.018 µg-Cr(VI)/cm^2 was recorded for this person, and he was not required to complete round three. Forty-nine of the 54 subjects did not respond to either concentration and therefore proceeded to round three.

Round Three. In round three, 1 of the 49 subjects responded to both 0.88 and 0.18 µg-Cr(VI)/cm^2. The MET for this person was recorded as 0.18 µg-Cr(VI)/cm^2. Also, 22 subjects responded to 0.88 µg-Cr(VI)/cm^2 but not 0.18 µg-Cr(VI)/cm^2, and therefore 0.88 µg-Cr(VI)/cm^2 was recorded as the MET for those subjects. Twenty-two of the volunteers tested in round three had no response to either concentration. Because these volunteers did not respond to any patch concentration less than the diagnostic patch, the MET recorded for them was 4.4 µg-Cr(VI)/cm^2. Table 3 presents the MET and cumulative frequency of responses for Cr(VI).

Statistical Analysis of Results

The data were analyzed by a computer modeling data technique (GENSTAT, 1982). A truncated log normal distribution was fitted to the data with maximum likelihood methods, which is a technique for choosing parameters of a selected distribution of the observed data such that the probability of response of the observed data is maximized (Rothman, 1987). The fit to the data was excellent as confirmed by the Chi-squared goodness of fit test, which gave a p value of >0.05. The 10% cumulative response MET for Cr(VI) was 0.089 µg/cm^2. The log normal distribution is conventionally used in the analysis of bioassay data and was used here (Armitage and Berry, 1987). Because a reaction at the maximum tested concentration was a criterion for inclusion in the study, the distribution of response concentrations is truncated. Thus, the truncated log normal distribution was fitted

TABLE 3
Cumulative Dermal Response of 54 Cr(VI)
Sensitized Participants to Various Concentrations of
Cr(VI)

Cr(VI) (µg/cm^2)	Minimum elicitation threshold response (%)	Cumulative response (%)
0.018	1/54 (2)	1/54 (2)
0.088	4/54 (7)	5/54 (9)
0.18	5/54 (9)	10/54 (19)
0.88	22/54 (41)	32/54 (59)
4.4	22/54 (41)	54/54 (100)

to the data set with the highest truncation point being the highest Cr(VI) concentration tested (4.4 μg-Cr(VI)/cm^2).

DISCUSSION

The results of this study corroborate earlier reports that Cr(VI)-sensitive people respond to serial dilutions of Cr(VI) in a fairly linear manner at moderate concentrations (Zelger, 1964, 1966; Burrows and Calnan, 1965; Anderson, 1960; Geiser *et al.*, 1960). Interestingly, almost half (22/54) of the Cr(VI)-sensitized volunteers in this study did not respond to Cr(VI) concentrations less than the diagnostic concentration of 4.4 μg-Cr(VI)/cm^2. Many of the subjects who failed to respond in round one had been thought by the physicians to be Cr(VI) sensitive due to earlier instances of positive responses to 0.5% $K_2Cr_2O_7$ in petroleum jelly or to 0.25% $K_2Cr_2O_7$ in TRUE-Test matrix. The 0.5% $K_2Cr_2O_7$ patch is no longer commonly used clinically in the U.S. due to the high rate of irritant (non-allergic) responses (Marks and DeLeo, 1992; Fischer and Maibach, 1989), although it is still used in Europe. Many dermatologists in North America believe that use of 0.25% $K_2Cr_2O_7$ to identify sensitization, as was done in this study, minimizes or eliminates the incidence of irritant reactions (Storrs *et al.*, 1989). Therefore, it is likely that many of those volunteers who initially reacted to 0.5% $K_2Cr_2O_7$ exhibited an irritant (false positive) reaction in earlier testing and that the 0.25% $K_2Cr_2O_7$ patch used in this study failed to produce the irritant response. Alternatively, it is possible that some of the subjects who failed to response to round one truly are Cr(VI) sensitive but have a threshold that is greater than 0.25% $K_2Cr_2O_7$. If this is true, then the results of the study could be considered "worst case" as only the hypersensitive volunteers were used to represent the sensitized population. The results of this study suggest that the 10% MET for Cr(VI)-induced ACD is approximately 0.089 μg-Cr(VI)/cm^2. No significant correlations between age, sex, or occupation were found at any of the tested concentrations.

Recently, Marks *et al.* (1996) demonstrated that the prevalence of Cr(VI)-related ACD in a large clinical population was approximately 1%. The prevalence rate in the general population is certain to be much less. Therefore, if Cr(VI)-sensitized subjects comprise less than 1.0% of the general population, an ACD-based soil clean up criterion that is protective of 90% of the Cr(VI)-sensitized population, will be protective of about 99.9% of the general population.

It is noteworthy that there is some uncertainty associated with reading a patch test response as an allergic vs. an irritant reaction. Nethercott (1990) has stated that the incidence of "misread" test sites can be as high as 20%. Because the patch test reactions in this study were interpreted by members of the NACDG, this rate should be much less as the interpretation of patch tests is a routine part of their practice. Because it is more common to misread irritant reactions as allergic than discount an allergic reaction as an irritant one (Fischer and Maibach, 1989), the

MET estimated from this study can be considered conservative (no less than predicted here).

Calculating an ACD-based Soil Concentration for Cr(VI)

Soil Concentration (mg - allergen/kg - soil)

$$= \frac{\text{MET}\left(\dfrac{\text{mg - allergen}}{\text{cm}^2 \text{ - skin}}\right) \times \text{CF}\left(\dfrac{10^6 \text{ mg - soil}}{\text{kg - soil}}\right)}{\text{SA}\left(\dfrac{\text{mg - soil}}{\text{cm}^2 \text{ - skin}}\right) \times \text{BVA (unitless)}}$$

As described by Horowitz and Finley (1994), an acceptable concentration of allergen in soil can be derived as above: where MET = the minimum elicitation threshold determined from patch test data; CF = conversion factor; SA = soil adherence factor; BVA = bioavailability. If the leachability (bioavailability) of the allergen from soil to skin is less than 100%, then the soil concentration must be adjusted upward accordingly.

The 10% MET for Cr(VI) in sensitized people, as identified in the patch testing study, was 0.089 μg-Cr(VI)/cm^2-skin. As discussed previously, a threshold derived from a 48-h, occluded patch test study is highly conservative (health protective) with respect to realistic environmental exposures. Accordingly, to remain consistent with USEPA's "reasonable maximal exposure" philosophy of risk assessment (wherein average and upper bound values are combined to give a "reasonable" estimate, rather than simply compounding worst case assumptions), it is appropriate to use an average value to represent soil adherence. The USEPA's suggested average value of 0.20 mg/soil cm^2-skin was used for this analysis.

With these values for MET and soil adherence factor, it can be estimated that a concentration of 445 mg Cr(VI)/kg soil (445 ppm) should not pose an ACD hazard. Hence, even if it were conservatively assumed that all Cr(VI) in soil adhering to skin was able to leach into skin moisture and cross the skin barrier, a soil concentration of 445 ppm Cr(VI) should protect the vast majority of the Cr(VI)-sensitized population from the ACD hazard.

CONCLUSION

This patch testing study was designed to identify a MET for Cr(VI)-induced ACD on a μg-Cr(VI)/cm^2-skin basis so that the results could be used to derive a safe soil concentration for persons with ACD. The results suggest that the 10% MET for

Cr(VI)-induced ACD is about 0.089 μg/cm², and that soil concentrations of approximately 450 ppm Cr(VI) or less do not produce ACD in a vast majority of the general population. This study indicates that traditional clinical patch testing, which is used routinely to diagnose sensitivity, can be modified so that the data can be used to resolve questions involving quantitative risk assessment.

CASE STUDY #2: Using Human Sweat to Extract Hexavalent Chromium from Chromite-Ore Processing Residue: Applications to Setting Health-Based Cleanup Levels

BACKGROUND

As described in Case Study #1, if one assumes 100% leachability of Cr(VI) from soil, then a soil concentration of approximately 450 ppm Cr(VI) would be protective of ACD in a vast majority of the general population. The actual degree to which Cr(VI) may leach from solid matrices into human sweat (and thus become potentially bioavailable) is not clearly understood. Wainman *et al.* (1994) performed a leachability study using deionized water and a synthetic sweat solution to extract COPR samples of different Cr(VI) concentrations [up to 700 ppm Cr(VI)] and particle sizes. They reported that up to 32 and 83% of the Cr(VI) present in fine COPR particles (size fraction d < 2 mm) was extracted into water and synthetic sweat, respectively. The New Jersey Department of Environmental Protection (NJDEP) and suggested that Cr(VI) present in the COPR could extract into human sweat to yield sweat concentrations that are equivalent to the initial COPR concentrations [i.e., 10 ppm Cr(VI) in COPR will yield 10 ppm Cr(VI) in sweat] (NJDEP, 1990). An ACD-based soil cleanup standard of 10 ppm Cr(VI) in New Jersey was then proposed on this basis (NJDEP, 1992a).

Similar to the circumstances described in Case Study #1, the appropriateness of the available data was somewhat lacking at the time the issue was initially raised. Specifically, the Wainman *et al.* (1994) study used deionized water and synthetic sweat rather than actual sweat. Synthetic sweat, which metallurgists have used to test corrosion resistance and nickel leaching from metals (Carter, 1982; Randin, 1987), is a mixture of sodium chloride, ammonium chloride, urea, lactic acid, and glacial acetic acid (Swiss Watch Industry Specification NIHS 96–10), whereas human sweat is a much more complex medium that also contains measurable amounts of oils, fatty acids, organic materials, chlorine, calcium, iron, magnesium, nitrogen, phosphorous, potassium, sodium, and sulfur, among many trace elements (International Commission on Radiological Protection [ICRP], 1984). Due to these and other differences, it is possible that the leaching capabilities of human sweat are quite different from water or synthetic sweat. Accordingly, the experiment described in this case study sought to provide the most accurate determination of the potential skin leachability of Cr(VI) from the COPR by using actual human sweat as the extractant. The experiment was designed to include a wide range of

Cr(VI) concentrations and sweat-to-COPR ratios in order to address a broad range of plausible conditions (including worst case) of environmental exposure to the COPR. The results of this study are directly applicable to setting ACD-based cleanup standards. Additional detail concerning the design, conduct, and interpretation of this study may be found in Horowitz and Finley (1993).

METHODS

Chromite-Ore Processing Residue Concentrations of Cr(VI)

Based on the range of Cr(VI) concentrations known to exist in the COPR, the following Cr(VI) concentration ranges were targeted for field collection: 5 to 20 ppm, 50 to 150 ppm, and >1000 ppm. The Cr(VI) concentrations in the samples used in this study are summarized in Table 4. Table 5 summarizes the pH, total organic carbon content, and moisture content of the samples.

Sweat:Chromite-Ore Processing Residue Ratios

Two sweat:COPR ratios were used to extract each of the samples. A 20:1 ratio was used to approximate perspiration conditions associated with moderate physical

TABLE 4
Concentrations of Hexavalent Chromium in COPR[a]

COPR Sample number	Targeted COPR concentration	Actual COPR concentration
1	5–20 ppm Cr(VI)	16 ppm (Cr(VI)
2	50–150 ppm Cr(VI)	136 ppm Cr(VI)
3	>1000 ppm Cr(VI)	1240 ppm Cr(VI)

[a] COPR, chromite ore processing residue.

TABLE 5
COPR Characteristics

COPR identification number	pH	Total Organic Carbon Content (%)	Percent moisture content (g water/g COPR)
1	9.1	1.5	14
2	7.5	3.5	31
3	8.4	5.2	41

activity; a 5:1 ratio was used to approximate conditions of resting or light physical activity. The 20:1 ratio was based on the observation that adults engaged in moderate physical activity will produce approximately 2 l sweat per day (ICRP, 1984). Using a total skin surface area for the average male adult of approximately 20,000 cm^2 (USEPA, 1989), yields a sweat rate of 0.0042 g sweat/cm^2-skin-h. For the purposes of this analysis, a conservative soil loading (adherence) factor of 0.001 g/cm^2 was used; this value is the maximum of the range of values recommended in the USEPA's *Dermal Exposure Assessment: Principles and Applications* (0.0002 to 0.001 g/cm^2) (USEPA, 1992). Using these estimates, a sweat: COPR ratio of 25:1 for a 6-h exposure period is calculated as:

$$\frac{SR * time}{SL} = ratio$$

where SR is the sweat rate (g sweat/cm^2-skin-h), time is the duration of soil contact on skin (h), and SL is the soil loading rate (g soil/cm^2-skin). Thus,

$$\frac{(0.0042 \text{ g sweat/cm}^2 \text{ skin - h}) * (6 \text{ h})}{0.001 \text{ g soil/cm}^2 \text{ skin}} = 25:1 \text{ (g sweat : g soil)}$$

In summary, if COPR adhered to skin in the morning hours, and one undertook moderate physical activity for a 6-h period, the total amount of sweat available to extract the COPR would yield a sweat:COPR ratio of approximately 25:1. This ratio was rounded to 20:1 for this study. For the resting state or light activity, a sweat:COPR ratio of 5:1 was determined to be a reasonable estimate (ICRP, 1984).

Chromite-Ore Processing Residue Preparation and Analysis

After collection, the COPR samples were first dried in an oven for 24 h at 105°C. They were then homogenized to create uniform samples. This process included sieving the samples to obtain a uniform particle diameter of ~500 µm. The use of such small particles in this study is intended to approximate the fine dusts that might adhere to skin as a result of environmental exposure. For example, Yang *et al.* (1989) have suggested that particles that adhere to the skin have an average diameter of <150 µm. As the relative surface area of the COPR particles is greatest in the "fines" fraction, using small particles ensures that the leaching potential of Cr(VI) from COPR is maximized. Following drying and homogenization, the samples were analyzed for Cr(VI) using USEPA Method 3060A/7196A.

Sweat Collection Methodology

Sweat was collected from seven young to middle-aged male adults working in Hudson County who voluntarily elected to participate in the study. Prior to sweat collection, the workers showered, washed, and donned modified Level C safety equipment. Level C equipment was used as it requires full-body protection, including a full face piece, full-body chemical-resistant suit (Tyvek), safety boots, gloves, and a hard hat, thereby enhancing sweat production. The workers engaged in moderate work activity for several hours. They were able to replenish their fluid loss by water intake as necessary, and activities were monitored by a health and safety officer. The volunteers wore minimal clothing under the Tyvek suit, such that most of the sweat ran down the body and collected in the safety boots. The sweat was collected by pouring the fluid that pooled in the boots into amber glass bottles. The samples were composited in the field prior to being sent to the laboratory for analysis.

Approximately 4 l sweat was collected from all seven volunteers. The sweat composite was analyzed for pH, total suspended solids, total organic carbon, oil and grease, and Cr(VI) and total chromium content. The pH of the unfiltered and filtered sweat was measured at 7.2 and 8.0, respectively, which indicates that the sweat composite was neutral to slightly basic. Analyses were performed after the sweat was filtered through 1.0-, 0.45-, and 0.1-μm filters (in series). Similar analyses were conducted on a sample of distilled water that had been placed in a boot (boot blank). The purpose of analyzing a boot blank was to account for any Cr(VI) extracted from the boots. The boot blank analyses were all below the detection limit of 0.010 ppm Cr(VI).

Three Cr(VI) spikes were added to the composited sweat at 0.5, 1.0, and 10 ppm to determine the recovery of Cr(VI) for quality control/quality assurance purposes. The percent recovery ranged from 84.5 to 101.3%, which meets the USEPA Contract Laboratory Program (CLP) accuracy criteria. The sweat was stored in tightly sealed containers at 4°C prior to the extractions.

Chromite-Ore-Processing Residue Extraction Methodology

Portions of 20 and 5 g of each COPR concentration were weighed and placed in 300-ml biochemical oxygen demand (BOD) bottles. One hundred milliliters of sweat was added to achieve sweat:COPR ratios of 5:1 and 20:1, respectively. Triple replicates of the 16 ppm Cr(VI) COPR samples were extracted and analyzed to assess method precision. Single extractions were performed on the other COPR samples. The bottles were covered tightly with a shaved glass tapered stopper and partially immersed in a water bath at 30°C (approximate mean skin temperature) (Pandolf et al., 1988; Kuno, 1934) for 12 to 16 h. The bottles were gently swirled once per hour (5 revolutions) during the last 3 h of extraction to mix the sweat with the COPR. The sweat:COPR mixtures were filtered following 12 to 16 h of extraction by suction filtration using three filters in series: 1.0-, 0.45-, and 0.1-μm filter. Each extract was collected in a 100-ml graduated cylinder, and the volume recovered from the extraction procedure was recorded and used to calculate the final concentration of Cr(VI) in the sweat.

Extraction of Cr(VI) by Sweat

Table 6 summarizes the results of Cr(VI) analyses of the sweat extracts. None of the extracts from the 16 ppm triple replicate samples had detectable levels of Cr(VI) at either the 5:1 or 20:1 sweat:COPR ratio (limit of detection approximately 0.010 ppm). Similarly, the 20:1 extract of the 136 ppm sample did not contain detectable Cr(VI) levels. The 5:1 extract of the 136 ppm sample contained 0.09% of the Cr(VI) initially present in the COPR, with a resulting sweat concentration of 0.027 ppm. Both the 5:1 and 20:1 extracts of the 1240 ppm Cr(VI) sample contained approximately 0.05% of the Cr(VI) initially present in the COPR, with resulting sweat concentrations of 0.133 and 0.040 ppm, respectively.

In summary, within the limitations of the analytical methods utilized: (1) Cr(VI) was not detected in sweat extracts of the 16 ppm Cr(VI) COPR at a limit of detection of 0.010 to 0.014 ppm; (2) the measured amount of Cr(VI) extracted from the 136 ppm and 1240 ppm samples did not exceed 0.09% of the total mass of Cr(VI) present in the COPR, and (3) final measured concentrations of Cr(VI) in sweat extracts from the 136 ppm and 1240 ppm samples were 0.133 ppm or less.

TABLE 6
Amount of Hexavalent Chromium
Extracted from COPR by Human Sweat

Homogenized Cr(VI) COPR concentration (ppm)	Sweat-to-COPR dilution ratio[a]	Cr(VI) concentration in extract (ppm)[b]	Percent of Cr(VI) extracted
16	5:1	ND (0.011)	ND
		ND (0.011)	
		ND (0.014)	
16	20:1	ND (0.010)	ND
		ND (0.010)	
		ND (0.011)	
136	5:1	0.027	0.09
136	20:1	ND (0.010)	ND
1240	5:1	0.133	0.05
1240	20:1	0.040	0.06

Note: ND, nondetectable.

[a] The dilution ratio was accomplished using 100 ml of sweat and either 20 g (5:1) or 5 g (20:1) of COPR. For the 1240 ppm Cr(VI) samples, 200 ml of sweat was used with either 40 g (5:12) or 10 g (20:1) of COPR.

[b] The numbers in parentheses represent the Cr(VI) limit of detection.

A few studies have attempted to approximate the amount of Cr(VI) that may be leached from the COPR in the presence of human sweat (e.g., Wainman *et al.*, 1994). However, these studies to date have employed alkaline buffer (USEPA Method 3060) or synthetic sweat as the extractant. The unique aspect of this study is that actual human sweat was used to solubilize the Cr(VI) from the COPR. These results have direct relevance to setting soil cleanup levels that are protective of ACD. Specifically, if the minimum elicitation threshold for Cr(VI) in soil is 450 ppm under conditions of 100% Cr(VI) extractability, and it is shown that 0.1% or less of Cr(VI) will actually leach from COPR in the presence of human sweat, then COPR concentrations much higher than 450 ppm Cr(VI) would be needed to elicit ACD after dermal contact. As described in Case Study #1, if 1% or less of the general population, including children, are Cr(VI) sensitive (Paustenbach *et al.*, 1992), an elicitation threshold at which no more than 10% of the sensitized population responds would protect >99.9% of the general population. In this analysis, the highest Cr(VI) concentration tested was 1240 ppm Cr(VI) in COPR. Thus, COPR concentrations up to 1240 ppm Cr(VI) are protective of ACD in at least 99.9% of the general population.

The extraction results reported in this article appear to be inconsistent with those of Wainman *et al.* (1994). Specifically, Wainman *et al.* (1994) reported that up to 32 and 83% of the Cr(VI) present in COPR was leached into deionized water and synthetic sweat, respectively, while the results of this study demonstrated that no more than 0.09% of the Cr(VI) in the COPR extracted into human sweat. These differences could be due to the fact that Wainman *et al.* (1994) extracted for 24 h at 39°C using constant agitation, while in this study the COPR was extracted for 12 to 16 h at 30°C using little agitation. However, it may also be that human sweat is a less-effective extracting medium or that organic constituents present in human sweat (but not synthetic sweat) reduce Cr(VI) to Cr(III). In addition, an extractant temperature of 30°C was used, which approximates skin temperature (Sulzberger and Herrmann, 1954), while Wainman *et al.* (1994) used a temperature of 39°C. Further, Wainman *et al.* (1994) agitated the solution/COPR mixture throughout the duration of extraction. Vigorous mixing is unlikely to be representative of particle/ sweat interactions on the skin surface and will leach more material from the COPR than would occur under normal exposure conditions. Occasional swirling of the extraction flask or very low agitation seems more representative of environmental conditions. In summary, the results reported herein would seem to be more applicable to setting ACD-based standards for COPR, as the study design most closely approximates environmental exposure conditions.

Sweat rates and sweat composition may vary somewhat with age, gender, and race. In general, men tend to perspire more than women (Rothman, 1954; Avellini *et al.*, 1980; Buono *et al.*, 1991), and young adults tend to perspire more than middle-aged individuals (Pandolf, 1991). Ethnic diversity apparently has no influence on sweat rate (Gibson and Shelley, 1948; Herrmann *et al.,* 1952; Sulzberger and Herrmann, 1954; Yousef *et al.,* 1984). The sweat rates used to develop the sweat:COPR ratios in this study were based primarily on values measured in young male subjects. As males perspire somewhat more than and females, and it is known that children sweat less than adults (Drinkwater and Horvath, 1979), these ratios should be considered conservative and applicable to the general population. The sweat rates used in this analysis were derived from whole-body perspiration studies and therefore represent an average sweat ratio. It is possible that the anatomical areas that are typically exposed to the environment (i.e., arms, hands, face) might have slightly higher or lower sweat rates than those used in this study. However, in general, sweat rates for different anatomical parts of the hands, chest, back, and forehead differ by less than an order of magnitude (Herrmann *et al.,* 1952), and therefore any such differences will have a trivial influence on selecting sweat to soil ratios. With respect to sweat composition, there are essentially no major differences due to age, race, or gender (Sulzberger and Herrmann, 1954), although women tend to have a lower sweat pH by 0.5 pH units (ICRP, 1984; Rothman, 1954). Hence, the sweat used in this study should also be considered applicable to the general population.

It is worth noting that the ratio of sweat to soil on the skin of a given individual is likely to vary throughout the day, depending on soil exposure conditions and degree of perspiration. In order to derive sweat:soil ratios for this analysis, it was assumed that the exposed skin became covered with soil after initial exposure in the morning and that the soil layer remained on the skin throughout the day. It was further assumed that sweat would extract this soil layer throughout the day. Hence, although sweat:soil ratios are likely to demonstrate a dynamic proportionality throughout the day, the protocol used here is a very conservative representation of environmental conditions.

It is possible that the degree of extraction is influenced by COPR characteristics such as pH, particle size, etc. The wide ranges of pH (7.5 to 9.1) and moisture content (14 to 41%) among these samples suggests that the extraction results are applicable to most, if not all, of the COPR in Hudson County. As noted earlier, "fines" were used to approximate worst-case extraction conditions. Hexavalent chromium generated from manufacturing processes other than chromite-ore processing might demonstrate different affinities for soil. Thus, it is critical to consider site-specific conditions when assessing Cr(VI)-related health risks or setting cleanup standards for Cr(VI) in soils.

CONCLUSION

This study, in combination with the patch-test results described in Study #1, indicates that COPR Cr(VI) concentrations up to 1240 ppm Cr(VI) will protect against the development of ACD in the general population, and in general ACD is unlikely to be a significant health concern for environmental exposures to the COPR in Hudson County. This conclusion is supported by the fact that there have been no substantiated instances of ACD resulting from exposure to the COPR in the environment (Fagliano and Savrin, 1994).

CASE STUDY #3: Urinary Excretion of Chromium Following Ingestion of COPR in Humans: Implications for Biomonitoring

BACKGROUND

Chromium that is systemically absorbed is excreted almost exclusively in the urine (WHO, 1988; Anderson *et al.*, 1982). Urinary excretion of chromium has often been used as a biomarker of exposure to elevated airborne chromium concentrations in occupations such as chrome-plating and welding (Tola *et al.*, 1977; Mutti *et al.*, 1979; Tassavainen *et al.*, 1980; Lauwerys, 1983; Welinder *et al.*, 1983; Minoia and Cavalleri, 1988; Lindberg and Vesterberg, 1989; Bonde and Christensen, 1991). Recently, several investigators have attempted to assess whether increased urinary chromium concentrations can be measured in individuals exposed to COPR in New Jersey soils (Bukowski *et al.*, 1992; Stern *et al.*, 1992; Fagliano, 1994). (See Background section of Case Study #1 for detail on the origin and characteristics of the COPR.) In the first study, Bukowski *et al.* (1992) found that the mean urinary chromium concentration in New Jersey park rangers working near COPR on state-owned land (0.60 µg-Cr/g creatinine) was no greater than that of a "control" park ranger population (0.63 µg-Cr/g creatinine). In the second study, Stern *et al.* (1992) found no significant difference in the mean urinary chromium concentrations in individuals residing in areas built on COPR-containing soil (0.22 µg-Cr/g creatinine) vs. control residents distant from the COPR sites (0.18 µg-Cr/g creatinine). Stern *et al.* (1992) reported that mean chromium concentrations in household dusts and exposed soils at the COPR sites (153 µg/g dust) were significantly greater than in the control areas (127 µg/g dust).

The New Jersey Department of Health (NJDOH) has recently conducted a large "screening" survey (the NJDOH Chromium Medical Surveillance Program or "CMSP") in which single urine samples were collected from approximately 2200 individuals living in New Jersey, where COPR was thought to have been used as fill. The chromium concentrations in those urine samples were compared with concentrations measured in a "control" group of approximately 300 individuals living in areas distant from any known source of COPR. Based on the results of the comparison of "control" and "exposed" groups, the NJDOH concluded that some individuals in the "exposed" group may have elevated urinary chromium concentrations.

Typically, biomonitoring surveys operate on the premise that sufficient exposure will yield a measurable and reproducible increase in the body burden of the chemical of interest. To date, there has been little scientifically compelling evidence to indicate that exposure to COPR will result in increased urinary chromium concentrations (i.e., Stern *et al.* [1992] and Bukowski *et al.* [1992]). More importantly, as recently noted by an expert panel that was convened specifically to review the CMSP protocol, it is unclear whether even exaggerated contact with the COPR material could result in increased urinary chromium concentrations (Anderson *et al.,* 1993). Specifically, the concentrations of airborne chromium measured at several COPR sites in New Jersey are far less than those required to cause increased urinary chromium concentrations (Gargas *et al.,* 1996), and therefore incidental ingestion of household dusts and soils is the only plausible pathway by which sufficient exposure could occur to influence urinary chromium concentrations. As discussed by the expert panel, however, even if one uses conservative assumptions regarding dust and soil ingestion rates, chromium concentrations in the COPR, and chromium bioavailability from the COPR, it is unlikely that urinary chromium concentrations distinguishable from "background" would be measured (Anderson *et al.,* 1993).

In this study, this uncertainty is addressed by measuring urinary chromium concentrations in humans following ingestion of the COPR material for 3 d at levels approximating those described in the Stern *et al.* (1992) study (designated the "low-dose" group here). Two other subjects ingested larger quantities of COPR ("high-dose" group) to evaluate a "worst-case" scenario. By evaluating the urinary chromium excretion profile in the volunteers before, during, and after COPR ingestion, it is possible to draw conclusions regarding the feasibility of urinary biomonitoring as a means to evaluate environmental exposure to COPR. Additional details regarding this work may be found in Gargas *et al.* (1994a).

METHODS

Soil Collection and Analysis

The soil and COPR mixture used in this study was obtained from a surficial (0- to 6-in.) soil sample collected at a former chromite-ore processing facility. The mixture was homogenized by standardized methods (ASTM, 1992) and passed through a no. 35 brass sieve. Soil particles equal to or less than 500 μm were collected and used to approximate the dusts that accumulate on household surfaces.

The mixed soil was sterilized in an autoclave at 15 psi and 120°C for 3 consecutive days, 8 h/d (24 h total). The autoclaved sample was cultured for aerobic and anaerobic organisms and fungi. The soils used in this study were negative in all cultures. Representative samples of the sieved, sterilized soil were extracted and analyzed for total chromium [Cr(VI) and Cr(III)] and Cr(VI) content

670

using USEPA Method 3050/6010 and a modified version of USEPA Method 3060A/7196A, respectively. The soil was also analyzed for all substances on the USEPA Target Compound List (TCL). The total chromium concentration was determined to be 103 ± 20 mg Cr/kg soil (mean $\pm$ SD; $n = 4$), and the Cr(VI) concentration was 9.3 ± 3.8 mg Cr(VI)/kg. A total chromium concentration of approximately 100 mg/kg soil was selected for ease of comparison of these results with those of previous studies (Stern *et al.*, 1992). A variety of TCL chemicals was also present at relatively low concentrations.

Dosing Regimen

The study lasted 6 d, beginning with the first morning void on day 1, until the first morning void on day 6. Capsules containing soil were ingested on each of days 2, 3, and 4. No capsules were ingested on day 1 in order to provide baseline excretion levels in the urine for each participant. Continued chromium elimination was assessed on day 5 and up to the first morning void on day 6. This time frame was considered sufficient to assess chromium elimination following the last dose as the major elimination half-life in humans has been reported to be about 15 to 41 h (Tassavainen *et al.*, 1980).

The "low-dose" volunteers ingested 400 mg of the soil on each of days 2, 3, and 4, for a dose of 41 μg chromium/d, and the "high-dose" volunteers ingested 2 g of soil for each of 3 d, for a dose of approximately 206 μg-Cr/d. A single bolus dose was used to approximate the "worst-case" environment setting, where a person might ingest a large amount of soil over a period of an hour or less. To have used several smaller doses would be more realistic but would lessen the ability to measure increased chromium in the urine. One participant served as a blind control and received capsules containing 400 mg of ground coffee (placebo).

Sample Collection/Urine Analysis

Participants collected urine samples separately at each voiding during the study. Each void was collected in a separate, polyethylene, leak-resistant container approved for environmental sampling (Nalge Company, Rochester, NY). All samples were stored at 4°C until being shipped to the laboratory in a cooler with ice packs. Chain of custody was maintained for all samples and approximately 10 ml of each individual sample was analyzed for total chromium, creatinine, and specific gravity as described previously (Stern *et al.*, 1992; Gargas *et al.*, 1994b; Veillon, 1982). The limit of detection (LOD) for chromium was 0.2 μg-Cr/l urine, or approximately 0.15 μg-Cr/g creatinine (assuming an average excretion of 1.3 g creatinine/l urine [Anderson *et al.*, 1993]). Quality assurance and quality control procedures were as described previously (Gargas *et al.*, 1994b).

Capsule Preparation

Four hundred milligrams of the prepared soil was weighed on a certified scale (Overhold's Pharmacy, Champion, OH) and placed in each of 18 separate gelatin capsules (Eli Lilly and Co., Indianapolis, IN; size 00). An empty capsule was analyzed for total chromium and was found to contain very small quantities (0.014 µg-Cr/capsule). Three capsules containing 400 mg of soil were provided to each of six volunteers in the low-dose group. One gram of soil was weighed and placed in each of 12 separate gelatin capsules for the high-dose study. One participant was not exposed to COPR but was provided three capsules containing 400 mg of ground coffee (placebo).

Human Use Committee

A Human Use Committee (Richard A. Anderson, Ph.D., Department of Agriculture; Edward J. Calabrese, Ph.D., University of Massachusetts; John Doull, M.D., University of Kansas Medical Center; and James R. Nethercott, M.D., Baltimore, Maryland) reviewed the study protocol for this experiment to assess whether the study was appropriately designed and sufficiently rigorous to make valid conclusions and to determine whether the experiment posed any health hazard to the participants. The panel concluded that the study protocol design was adequate to satisfy the proposed objectives and that based on a quantitative evaluation of the doses associated with ingestion of the chromium and TCL compounds in the COPR, the study posed no significant risk of injury to the participants.

Human Volunteer Description

All volunteers reviewed the study protocols, analytical soil results, and Human Use Committee comments and signed an informed consent form prior to participation in this study. All participants were required to complete a questionnaire to provide basic physiological and medical information. No unusual medical conditions were reported, and all individuals were found to be eligible to participate in this study. Participants were instructed to follow a normal diet and to keep a detailed diary of all food and beverage intake. The five male participants weighed between 170 and 210 lb and were between 29 and 51 years old, and the four female participants weighed between 107 and 130 lb and were between 23 and 40 years old. The participants live in various regions of the U.S., and none live close to areas known to contain COPR.

RESULTS

Each of the volunteers excreted an average of approximately five voids per day; a total of 228 samples was collected from the placebo, low-, and high-dose groups.

During "baseline" conditions (day 1), six of the nine subjects had no detectable urinary chromium (average limit of detection per sample, approximately 0.15 μg-Cr/g creatinine). The other three participants (H4, H5, and H8) had urinary background chromium concentrations ranging from nondetectable (7 of 13 samples) to measurable levels ranging between 0.08 and 1.4 μg-Cr/g creatinine.

In the low-dose study, there were very few samples (8 of 125) collected during the dosing period (days 2 to 4) that contained measurable levels of chromium. Subject H4 had no detectable urinary chromium throughout the course of the dosing. The eight samples with measurable chromium concentrations ranged from 0.14 to 1.9 μg-Cr/g creatinine and appeared randomly, with no relationship to daily or cumulative dose. For each individual, the mean chromium concentration of the "baseline" samples was not significantly different from the mean chromium concentration of the samples collected during days 2 to 4 (data not shown). Subject H7 ingested the placebos and had no measurable urinary chromium.

In the high-dose study, one of the two subjects (H9) had no measurable urinary chromium throughout the course of the dosing (average LOD approximately 0.15 μg-Cr/g creatinine). As shown in Table 7, high-dose Subject H8 had measurable chromium levels throughout the baseline day and during the days of dosing. The highest chromium concentration measured in a sample from H8 (1.4 μg-Cr/g creatinine) occurred during baseline conditions. The data for Subject H8 are provided to illustrate the type of information collected for each participant. Subject H8 had background chromium levels clearly elevated above those of the other participants. The study log for Subject H8 indicates a high level of physical activity, which may be attributable to the elevated chromium concentrations. Specifically, it is well established that physical exertion accompanied by heavy sweating and subsequent hydration can cause increased urinary chromium levels (Bukowski *et al.,* 1991). As with the low-dose individuals, there was no significant difference between "baseline" urinary chromium concentrations and those measured during dosing. In addition, the appearance of chromium in the urine of Subject H8 was random with respect to daily and cumulative dosing.

DISCUSSION

The background urinary chromium concentrations measured in this study (<0.2–2.9 μg/l) compare well with the range of background chromium concentrations reported by IARC (1990) for healthy individuals (0.24 to 1.8 μg/l) and for these same volunteers from an earlier study (<0.2 to 1.3 μg/l) (Gargas *et al.,* 1994b). These results for typical background urinary chromium are also comparable to those found for a group of 76 adult males and females who participated in a clinical study (mean = 0.20 ± 0.01 μg/l; range = 0.05 to 0.58 μg/l [Anderson *et al.,* 1982]) conducted in Maryland. The wide range of background concentrations found in this study and in previous studies highlights the inherent difficulty in attempting

TABLE 7
Volume, Specific Gravity, Creatinine, and Total Chromium Measured in the Urine of Volunteer H8 (High-Dose Study)

Day of study	Time of specimen collection	Volume (ml)	Specific gravity	Creatinine (mg/l)	Total chromium (μg/l)	Total chromium (μg/g creat.)	Total chromium excretion (μg/d)
1	0730	280	1.025	1688	0.3	0.18	
	1438	220	1.027	2656	0.5	0.19	
	2142	205	1.029	3047	ND[a]	ND	
	0045	140	1.028	2029	2.9	1.4	0.60
2	0935	405	1.023	1809	0.5	0.28	
	1525	240	1.026	2185	0.7	0.32	
	2230	145	1.031	4412	0.7	0.16	0.47
3	0640	210	1.031	3278	0.5	0.15	
	1155	225	1.016	1088	ND	ND	
	1650	390	1.023	1848	ND	ND	
	1940	345	1.01	672	ND	ND	
	2110	285	1.007	552	0.2	0.36	0.16
4	0535	300	1.025	2318	0.5	0.22	
	0950	235	1.025	1611	ND	ND	
	1500	295	1.023	1300	ND	ND	
	2050	440	1.02	1224	0.2	0.16	0.24
5	0555	395	1.025	1921	0.2	0.1	
	0925	270	1.028	1133	ND	ND	
	1150	275	1.017	691	ND	ND	
	1325	265	1.01	533	ND	ND	
	1645	510	1.012	849	ND	ND	
	1900	250	1.02	1173	ND	ND	
	2035	380	1.008	467	ND	ND	
	2200	195	1.017	877	ND	ND	
	2400	170	1.025	1345	0.2	0.15	0.11
6	0510	350	1.024	1273	ND	ND	ND

[a] None detected (Cr limit of detection, 0.2 μg/l or 0.2 μg/l divided by g creatinine/l).

to discern normal "background" chromium excretion (from diet, smoking, etc.) from other potential chromium sources.

The individual postdosing urinary chromium excretion periods did not demonstrate dose-dependent patterns. Indeed, there was no significant difference between pre- and postdosing mean urinary chromium concentrations in either the low- or the high-dose groups (Table 8). Further, there was no difference in mean urinary chromium concentrations between the low- and the high-dose groups on any given day (Table 8). This relationship also holds true of the data collected for each individual. Hence, in no case did ingestion of the COPR material result in increased urinary chromium concentrations.

The results of this study are consistent with those of Bukowski *et al.* (1992) and Stern *et al.* (1992), both of which showed no significant difference in mean urinary

TABLE 8
Mean Daily Spot
Urine Chromium Concentrations

Day	Group (μg-Cr/g creatinine)	
	Low-Dose (n = 6)	High-Dose (n = 2)
1 (background)	0.14 ± 0.20 (26)[a]	0.26 ± 0.42 (10)
2 (dose day)	0.18 ± 0.33 (27)	0.19 ± 0.12 (5)
3 (dose day)	0.19 ± 0.35 (31)	0.14 ± 0.09 (12)
4 (dose day)	0.14 ± 0.14 (30)	0.13 ± 0.07 (9)
5	0.11 ± 0.10 (31)	0.11 ± 0.05 (13)

[a] Mean ± SD (n samples); nondetects were assigned values at one-half the limit of detection.

chromium concentrations in groups of individuals exposed to COPR vs. control groups that were not exposed. The study described here is a refinement of these earlier studies because each individual served as his or her own control. The particle sizes used in this study (<500 μm) are thought to be representative of the fine particles that accumulate on household surfaces. Wainman *et al.* (1994) found less than a two-fold change in the amount of water extractable Cr(VI) from soils originating at COPR sites when the particle sizes ranged from <75 μm to <2000 μm. Therefore, even if household particle sizes are somewhat smaller or larger than those used in this study, the oral bioavailability of the chromium on the particle should not be significantly different. The biomonitoring survey recently conducted by NJDOH reported that household dust concentrations of total chromium ranged from 228 to 421 mg/kg. The soil concentration of 100 mg/kg was used in these studies to be consistent with the mean household dust concentration reported in the earlier study by Stern *et al.* (1992). However, because it was recognized that chromium levels in COPR are heterogeneous, very high doses of the 100 mg/kg material (up to 2000 mg of soil per day) were used to account for variability in dust concentrations.

Using the USEPA default adult soil ingestion rate of 100 mg/d and assuming all soil ingestion throughout the day comes from indoor dust (a very conservative assumption), an adult living with 400 mg/kg Cr in house dust would ingest 40 μg/d of Cr. The low-dose group ingested 41 μg-Cr/d and the high-dose group ingested 206 μg-Cr/d, all in a single dose (instead of many smaller doses taken throughout the day that would have been more realistic). Hence, the doses used in the low dose group are comparable to the plausible upper-bound doses that may have been experienced by similarly exposed residents (under worst-case situations). Further, the high-dose group surely received doses far in excess of what would be expected on a routine basis, probably more than ten-fold greater than expected for the vast majority.

This study shows that, even when using a sensitive dosing and sampling protocol, direct ingestion of gram quantities of fine COPR particles containing over 100 mg Cr/kg (up to 200 µg-Cr/d) does not influence urinary chromium levels. When one considers that exposure to soils and household dust typically occurs as a result of low-level incidental contact throughout the day (rather than a single bolus dose), it can be concluded that daily residential exposure to COPR levels much higher than 100 mg Cr/kg would also be unlikely to influence urinary chromium.

This study is the first in which human volunteers have ingested prescribed amounts of soil in order to assess chemical uptake from an environmental medium. The study involved no risk to the participants and was useful as a tool to test directly the merits of an ongoing environmental surveillance program. When planning or conducting health surveys, simple noninvasive techniques such as these can be used to examine the relationship between exposure to an environmental chemical and the associated body burden.

CASE STUDY #4: Human Ingestion of Cr(VI) in Drinking Water: Pharmacokinetics Following Repeated Exposure

BACKGROUND

Although Cr(VI) is currently classified by the USEPA as an inhalation carcinogen and can cause ACD following dermal contact, oral exposure to Cr(VI) has not been reported to cause health effects in humans or animals except at high doses (De Flora and Wetterhahn, 1989). The lack of oral toxicity at lower doses is believed to be due, in part, to the fact that the reductive conditions of the stomach convert ingested Cr(VI) to Cr(III) prior to systemic absorption (De Flora *et al.*, 1987; Donaldson and Barreras, 1966). De Flora *et al.* (1987) have shown, based on *in vitro* studies, that the reductive capacity of gastric juices in humans is likely to be on the order of tens of milligrams of Cr(VI) per day (De Flora *et al.*, 1987). Indeed, the USEPA cited De Flora *et al.*'s (1987) findings as one of the bases for raising the maximum contaminant level (MCL) of Cr(VI) in drinking water from 0.05 to 0.10 mg/l (USEPA, 1991).

Urinary excretion of chromium following ingestion of elevated, but safe doses of Cr(VI) has been examined recently (Finley *et al.*, 1996). Specifically, urinary chromium levels were measured in six individuals who ingested the equivalent of the USEPA's Cr(VI) reference dose (RfD; 0.005 mg/kg/d) for 3 d. Urinary chromium excretion rates and urinary chromium concentrations were significantly elevated above background in all individuals. The work described in Case Study #4 continues the examination of the pharmacokinetics and potential health effects associated with Cr(VI) ingestion in humans. Absorbed chromium is excreted primarily in the urine but, while elevated urine chromium concentrations can be used as a general marker of exposure to chromium, urine data alone do not give any definitive insight as to the valence state of the absorbed chromium [i.e., Cr(VI) vs. Cr(III)]. This is because any Cr(VI) that is systemically absorbed is reduced to Cr(III) *in vivo* (by reducing components of the blood and tissues) prior to urinary excretion (Lewalter *et al.,* 1985; De Flora and Wetterhahn, 1989). Hence, in the Finley *et al.* (1996) Cr(VI) RfD study and in other studies that specifically examine urinary chromium, it is difficult to draw conclusions concerning the valence state, and therefore the potential toxicity of the absorbed chromium.

Post-exposure analysis of blood components allows for such a distinction because of the difference in behavior of Cr(VI) vs. Cr(III) in the blood compartment *in vivo*. Specifically, it has been established that Cr(VI) easily enters blood cells and becomes permanently bound to cellular components, while Cr(III) enters blood cells to a lesser degree and does not become bound (Lewalter *et al.,* 1985; Wiegand *et al.,* 1985). Indeed, because Cr(VI) will remain bound to the red blood cell for the entire cell lifespan (approximately 115 to 120 d), radiolabeled Cr(VI) has been used as a medical research tool for years to examine the lifespan of red blood cells and blood volumes (e.g., Gray and Sterling, 1950, Aaseth *et al.,* 1982; Jacobs *et al.,* 1994). *In vitro*, it has been shown that when Cr(VI) is added to whole blood (Lewalter *et al.,* 1985), the chromium that becomes associated with the blood cells cannot be removed even after numerous dialyses. Conversely, dialysis quickly and completely removes chromium from blood cells spiked with Cr(III) chloride. These *in vitro* findings are consistent with the *in vivo* findings of Coogan *et al.* (1991). Coogan *et al.* (1991) dosed rats intravenously with Cr(VI) and found that RBC chromium levels were increased significantly 1 h post-dosing and that these elevated levels had not decreased 7 d later. When the animals were dosed orally with Cr(VI), RBC chromium levels were increased at the 1 h time point but had returned almost to background after 7 d. One hypothesis offered by Coogan *et al.* (1991) for the differences observed via these two routes of exposure was that the Cr(VI) was reduced in the gastric juices, followed by systemic absorption of a chromium form that is reversibly taken up by the RBCs. If this hypothesis is correct, one would be able to distinguish whether systemic uptake of Cr(VI) or Cr(III) has occurred by measuring post-exposure chromium concentrations of blood cells over time: a sustained plateau of elevated RBC chromium concentrations (i.e., irreversible binding) would be indicative of systemic Cr(VI) absorption, while an increase followed by a rapid decrease after cessation of dosing (a reversible interaction) would be indicative of Cr(III) absorption.

No investigation to date has ever definitively examined, vis-a-vis a blood compartment profile, whether oral doses of Cr(VI) in humans are truly reduced to Cr(III) prior to systemic absorption. In this study, we examine the dose-related pharmacokinetics of chromium *in vivo* following oral exposure to a series of Cr(VI) concentrations in drinking water (0.10 to 10.0 mg/l). Urine, red blood cell, and plasma chromium concentrations are measured at several time points pre- and post-dosing. The purpose of this study is to examine: (1) the merit of the USEPA's position that consumption of the drinking water standard of 0.1 mg Cr(VI)/l is safe because only Cr(III) is systemically absorbed, (2) the likelihood that complete gastric reduction occurs at much higher concentrations, including the maximally plausible concentrations that could be associated with chronic oral exposure to contaminated tapwater (defined herein as approximately 2 to 10 mg/l), and (3) identification of a maximum Cr(VI) dose that does not result in significant systemic chromium uptake. Additional detail regarding the design of this study may be found in Finley *et al.* (1997).

METHODS

Human Use Committee

A Human Use Committee comprised of three occupational physicians and one toxicologist (each a university faculty member) with experience in chromium toxicology reviewed the protocol prior to the study. The committee concluded that the study design was adequate to satisfy the objectives and that the doses were unlikely to pose health risks to the participants. The committee also noted that the participants were properly informed and advised of reported adverse health effects associated with chromium exposure.

Human Volunteer Description

Five healthy male Caucasian volunteers between 190 and 220 pounds and between the ages of 30 and 54 participated in this study. Prior to participation in the study, all individuals completed an informed consent and release form, as well as a questionnaire regarding basic physiological and medical information. No medical conditions that would preclude participation in the study were reported, and all participants were found to be eligible for the study.

Cr(VI) Drinking Water Preparation

Potassium chromate (K_2CrO_4; Aldrich Chemicals) was used as a source of soluble Cr(VI). The K_2CrO_4 and deionized water mixture was prepared for each participant using an analytical balance precise to 0.01 mg (Mettler, Model AE 163). All drinking water samples were prepared in 1 l increments and divided into three separate 333-ml portions. Each portion was placed in labeled 500-ml polyethylene bottles. The deionized water used in the preparation of the solution was analyzed and shown to contain no detectable levels of hexavalent chromium (<0.005 mg/l).

Analysis of a duplicate water sample containing 1.0 mg Cr(VI)/l as K_2CrO_4 was performed to assure the stability of Cr(VI) in the mixture for the duration of the testing period. The sample proved to be stable with no measurable reduction of Cr(VI) occurring over a period of 4 weeks (beyond the duration of the study). K_2CrO_4 was used to make five solution concentrations: 0.1, 0.5, 1.0, 5.0, and 10.0 mg Cr(VI)/l. Lancaster Laboratories (Lancaster, Pennsylvania) verified the concentrations of each solution using USEPA Method 7196. Water containing 5.0 and 10.0 mg Cr(VI)/l exhibited a distinct bright yellow color, while the 0.1 to 1.0 mg/l solutions were difficult to distinguish from typical tapwater.

Record Keeping and Dosing Regimens

Diet was not controlled in this study; the participants followed their normal dietary food and liquid intake, with the exception that they were prohibited from ingesting vitamin supplements containing vitamin C or chromium. Because many foods contain varying amounts of chromium, a detailed log of each participant's beverage and food intake was maintained throughout the study. The participants also recorded any physical activity (exercise) other than normal daily activity as well as ingestion of vitamin supplements or medications.

Participants ingested five different concentrations of Cr(VI) over the duration of the study (0.1, 0.5, 1.0, 5.0, and 10.0 mg Cr(VI)/l). One liter of each concentration was consumed so that the ingested daily doses were 100, 500, 1000, 5000, and 10,000 µg-Cr(VI)/d (respectively). Each dose was ingested for 3 consecutive days. A daily dose consisted of a 333-ml volume consumed at approximately 10:00 am, 3:00 pm, and 8:00 pm. A "no-dosing" period of at least 1 d was observed between consumption of the different concentrations.

Collection of Urine Samples

Each urine void was collected starting from day 1 (beginning with the first morning void) through the last day of study (including the last void before bedtime). Each sample was collected in a separate, labeled 500-ml polyethylene container. Labels provided the following information: study participant number (#1, #2, #3, #4, or #5), study day, number of void for the day (#1 for the first void of the day and all others numbered consecutively), and date and time of collection. Samples were stored at refrigerator temperatures until packed for shipment.

Collection of Blood Samples

The participants were assigned (according to their geographic location) to a medical clinic that collected the blood samples for analysis. The sample vials used in this study (glass vacutainer tubes with royal blue tops made by Becton Dickinson or Sherwood Medical) contained an ethylenediamine tetra-acetate (EDTA) anticoagulant and a guaranteed low metals content. The negligible concentration of chromium was validated through separate testing (contribution <0.4 µg-Cr/l). The labeled blood samples were sent to the laboratory via overnight delivery and were analyzed for chromium content within 48 h of collection.

Each individual had clinical screening tests performed before the 5 mg Cr(VI)/l and after the 10 mg Cr(VI)/l dose, including a complete blood count (CBC), clinical chemistries (SMA–20), and urinalysis. The purpose of these clinical tests was to screen for any significant alterations in hematology, blood chemistries, or urinalysis that might have been due to Cr(VI) ingestion.

Laboratory Methods and Quality Control

Urine Analyses. All urine samples were analyzed at NMS Laboratories (Willow Grove, PA). Total urine volume, specific gravity, and creatinine were measured. An Atago Uricon-*N*-urine-specific gravity refractometer (NSG Precision Cells; Farmingdale, New York) was used to determine urine specific gravities. An alkaline picrate procedure (Instrumentation Laboratory Monarch 2000, Instrumentation Laboratory, Lexington, MA) was used to measure urinary creatinine. Urinary creatinine was used to correct for urine volume differences between samples due to variable fluid intakes by the participants. Urine total chromium levels were measured using a PE–4100 ZL graphite furnace atomic absorption spectrometer (AAS). The limit of detection was 0.5 µg-Cr/l urine. The concentrations of total chromium in the samples were then obtained from a calibration curve of total chromium in urine.

Laboratory quality assurance and quality control consisted of laboratory spikes, laboratory controls (e.g., duplicate samples), and water blanks. Laboratory controls were obtained from Bio-Rad Laboratories for total chromium and creatinine. Calibration curves were prepared daily by testing a spiked sample in the matrix and controls were run after every ten samples. All results were found to be within acceptable quality control limits for the assay (within 10 to 15% of certified values).

Blood Analyses. All plasma and red blood cell samples were analyzed at NMS Laboratories using a PE–4100 ZL graphite furnace atomic absorption spectrometer (AAS). Samples were received by the laboratory already centrifuged in the vacutainer collection tubes. The plasma and red blood cell samples were treated with a chemical modifier prior to analysis with the AAS. The modifier consisted of deionized water, magnesium nitrate, nitric acid, and Triton X–100. The limit of detection was 0.3 to 0.5 µg-Cr/l for plasma samples and 2 µg-Cr/l for red blood cell samples.

Laboratory quality assurance and quality control included laboratory blanks and laboratory controls. The laboratory equipment was calibrated on a daily basis using plasma samples spiked with known chromium concentrations and UTAK serum with known background chromium levels. All results were found to be within acceptable quality control limits for the assay (within 10 to 15% of defined standards).

RESULTS

Due to scheduling conflicts, Subjects #2 and #3 did not continue the study after the 500 µg-Cr(VI)/d and 1000 µg-Cr(VI)/d doses, respectively. All other subjects completed all five doses of the study, but some of the blood samples were lost, spilled, clotted, or analyzed incorrectly at the laboratory. These missing data may limit interpretations to some degree, but the overall trends for blood chromium content across doses appears to be generally consistent.

Urinary Chromium Excretion

Urinary chromium excretion for each dose was calculated as the total amount of chromium excreted during the three dosing days and 1 d post-dose (4 d total). As shown in Table 9, each of the five subjects demonstrated an increase in the amount of urinary chromium excreted as a function of dose, beginning at the lowest dose (100 µg-Cr(VI)/d). Interindividual differences in the amounts of urinary chromium excreted by Subjects #2 to #5 were small (typically within a factor of 2 to 3 of one another) at any given dose, while Subject #1 excreted chromium at a rate that exceeded all the others, in some cases by over an order of magnitude. Further, while Subjects #2 to #5 excreted <2% of administered chromium at each dose (range of 0.2 to 1.8% for 4-d excretion), Subject #1 excreted > 2% of administered chromium at all doses (8% maximum, at the 10,000 µg-Cr(VI)/d dose). Subjects #2, #3, and #5 did not demonstrate a dose-related increase or decrease in percent of administered chromium excreted in the urine. Subject #1 also excreted a fairly consistent percentage of administered dose at the first four doses (2.7 to 4.9%), but may have experienced a dose-related increase at the highest dose (8%), and Subject #4 demonstrated a slight but consistent dose-dependent increase. It should be noted that urinary chromium excretion likely continued beyond the collection period, particularly at the higher doses, and that the "% excreted" values given here may underestimate the total percent of administered dose excreted in the urine.

Chromium Concentrations in Plasma

As shown in Figures 1 and 2, Subjects #4 and #5 demonstrated evidence of a dose-related increase in plasma chromium beginning at the 5000 µg-Cr(VI)/d dose,

681

TABLE 9
Urinary Excretion of Total Chromium over
4 Days Following Oral Ingestion of Cr(VI) in Drinking Water[a]

| Subject no. | 0 µg/d[c] | Administered Cr(VI) dose[b] | | | | |
		100 µg/d	500 µg/d	1,000 µg/d	5,000 µg/d	10,000 µg/d
#1	0.6 (pre-study mean)	8.8	54.4	134.8	402.2	2407
	1.5 (historical mean)	(2.9%)	(3.6%)	(4.5%)	(2.7%)	(8.0%)
	(0.3 – 2.6) (historical range)					
#2	0.3 (pre-study mean)	3.1	8.8	NA	NA	NA
	0.6 (historical mean)	(1.0%)	(0.6%)			
	(0.3 – 1.1) (historical range)					
#3	0.5 (pre-study mean)	0.9	12.7	15.1	NA	NA
	0.5 (historical mean)	(0.3%)	(0.8%)	(0.5%)		
	(0.4 – 0.6) (historical range)					
#4	0.3 (pre-study mean)	0.7	4.9	9.7	80.7	451.1
	0.5 (historical mean)	(0.2%)	(0.3%)	(0.3%)	(0.5%)	(1.5%)
	(0.3 – 1.4) (historical range)					
#5	0.4 (pre-study mean)	2.9	7.7	7.3	273.2	251.6
	0.6 (historical mean)	(1.0%)	(0.5%)	(0.2%)	(1.8%)	(0.8%)
	(0.2 – 0.9) (historical range)					

682

Arithmetic Mean	0.5 (pre-study mean)	3.3 (1.7%)	17.7 (1.2%)	41.7 (1.4%)	252.0 (1.7%)	1037 (3.5%)
	0.7 (historical mean)					
	(0.3 – 1.3) (historical range)					

Note: NA Not assessed for that specific subject.

a The microgram excreted values are the total urinary chromium excreted for each of the 3 dosing days and 1 d post-dose. Values in parentheses represent percent of administered chromium dose excreted within 4 days after the first dose. Percent chromium excreted was calculated by summing the total chromium in urine per day for 3 dosing days and 1 day post-dose, after correcting for pre-study background urinary chromium concentration, and dividing by the total amount of chromium administered.

b Cr(VI) dosages are 3 daily doses of 11 of 0.10, 0.50, 1.0, 5.0, and 10.0 mg Cr(VI)/l.

c Pre-study, historical, and the historical range from measurements taken prior to dosing. Historical values include background measurements from studies in addition to the current one that the volunteer has participated in.

while the plasma chromium concentrations of Subjects #2 and #3 did not clearly increase above baseline at any of the doses tested in these individuals (up to 500 and 1000 μg-Cr(VI)/d, respectively). Subject #1 exhibited a dose-related increase beginning at the 1000 μg-Cr(VI)/d dose. As shown in Figure 2, the increases in plasma chromium observed in Subjects #1 and #4 at the highest dose were followed by a rapid decrease after the last dose. Both subjects exhibited an initial 'rapid' decline wherein plasma chromium concentrations decreased by approximately half within 7 d, followed by a slower, gradual decline. Plasma concentrations in Subject #1 were still clearly elevated (> 20 μg/l) 14 d after the last dose.

Chromium Concentrations in Red Blood Cells

As shown in Figures 1 and 2, the RBC chromium profiles generally mirrored the plasma chromium profiles for each subject. There was no evidence of a dose-related increase in RBC chromium in Subjects #2 to #5 at doses ranging from 100 μg-Cr(VI)/d to 1000 μg/Cr(VI)/d. The interpretation of this part of the profile is admittedly limited by the fact that it was not possible to collect data throughout the three doses for these subjects. Subject #1 experienced a dose-related increase beginning at the 1000 μg-Cr(VI)/d dose, while Subject #4 did not exhibit an increase until the 10,000 μg-Cr(VI)/d dose. Subject #5 did not experience a clear dose-related increase through any of the five doses. Similar to the urinary and plasma data, Subject #1 experienced an elevation in RBC chromium (65 μg/l maximum) that was much greater than for the other subjects (e.g., 19 and 8.1 μg/l maximum for Subjects #4 and #5, respectively). As illustrated in Figure 1, the RBC chromium levels in Subjects #1 and #4 began to decrease within days following the 10,000 μg-Cr(VI)/d dosing period and exhibited a 'rapid' and then 'slow' decline similar to that observed in the plasma.

Clinical Findings

All of the clinical findings were negative. There were no appreciable changes in the measurements from pre- to post-dosing for any of the clinical parameters measured in the CBC, SMA–20, or urinalysis for any of the five participants. Also, no outward symptoms such as gastric upset or nausea were reported.

DISCUSSION

This study has identified several interesting features of chromium pharmacokinetics that, in addition to refining our basic understanding of chromium uptake and disposition in humans, can be applied to interpretation of biomonitoring surveys

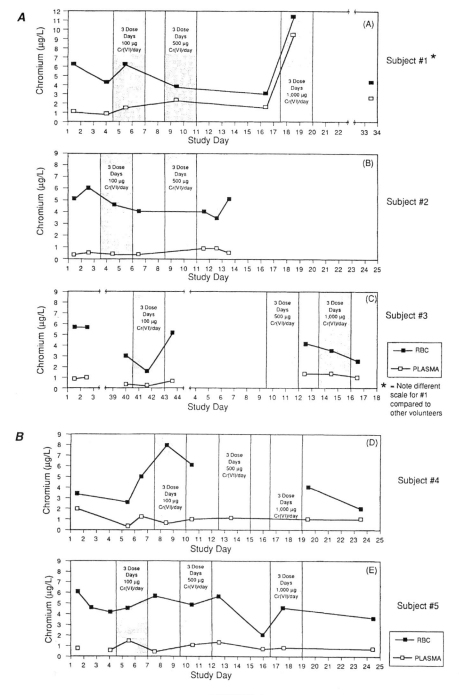

FIGURE 1

Plasma and RBCs of volunteers in concentrations of chromium following oral ingestion of 1 l/d of water containing either 0.1, 0.5, or 1 mg Cr(VI). At each concentration, the volunteers drank a total of 1 l of contaminated water per day for 3 consecutive days (shaded area represents days of dosing).

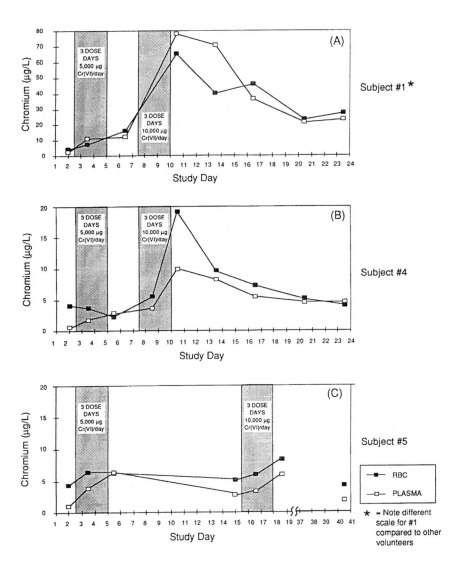

FIGURE 2

Plasma RBCs of volunteers in concentrations of chromium following oral ingestion of 1 l/d of water containing either 5 or 10 mg Cr(VI). At each concentration, the volunteers drank a total of 1 l of contaminated water per day for 3 consecutive days (shaded area represents days of dosing).

and other health risk assessment endeavors. First, it is clear that systemic chromium absorption occurs after ingestion of the USEPA's maximum contaminant level (MCL) of 0.10 mg/l as Cr(VI), as evidenced by the fact that all subjects had increased urinary chromium excretion during 3 d ingestion of 1 l of 0.10 mg Cr(VI)/l (100 µg-Cr(VI)/d). This is consistent with the findings of Finley *et al.* (1996), who demonstrated that ingestion of Cr(VI) at a dose equivalent to the

USEPA's Cr(VI) reference dose (RfD) of 0.005 mg Cr(VI)/kg-d (from which the MCL is derived) yielded elevated urinary chromium concentrations and excretion rates. The current study may be more relevant to environmental conditions than Finley *et al.* (1996), because in the present study the Cr(VI) was ingested in water throughout the day at a consumption rate that is consistent with residential tapwater consumption rates (1 l/d), while Finley *et al.* (1996) administered the Cr(VI) as a single bolus dose of finely divided $K_2Cr_2O_7$ packed into gelatin capsules. In short, it is clear that consumption of "safe" doses of Cr(VI) under environmental conditions (e.g., contaminated tapwater) will result in measurable increases in chromium excretion.

This finding has implications concerning the use of urinary chromium as a marker of environmental exposure. Urinary chromium concentrations are often used as biomarkers of exposure due to ease of collection and analysis and because systemically absorbed chromium is excreted primarily in the urine. However, as we have shown here, an excursion above "background" urinary chromium concentrations after oral exposure to Cr(VI), as measured intraindividually or by a comparison of "exposed" vs. "control" groups, is not sufficient evidence to conclude that an adverse health effect has resulted. Indeed, in the present study chromium excretion rates at the safe dose of 100 µg-Cr(VI)/d exceeded background chromium excretion rates by up to ten-fold. Although analysis of dietary uptake and other background exposures to chromium was not the focus of the current study, the historical background data for our volunteers indicate a fairly consistent central tendency and a limited range for background contributions using 24-h excretion measurements. For a further review regarding the use of urinary chromium as a biomonitoring tool see Anderson *et al.* (1993).

Second, it is important to note that dose-related increases in plasma and RBC chromium concentrations were not apparent following ingestion of the USEPA's MCL of 0.10 mg Cr(VI)/l, and that some subjects did not demonstrate a clear increase in blood chromium levels until water concentrations of up to 5 to 10 mg Cr(VI)/l were ingested (5000 to 10,000 µg-Cr(VI)/d). The RBC chromium concentration measured in Subject #4 during 100 µg-Cr(VI)/d dosing (8 µg/l) is higher than the pre- or post-dose RBC chromium concentrations. However, this is likely to be a function of variability at background and not related to chromium ingestion because the RBC chromium concentrations measured in Subject #4 during the 5000 and 10,000 µg-Cr(VI)/d doses were less than 8 µg/l. The apparent inconsistency between the urine and blood data is probably a result of the fact that blood samples were collected in an intermittent, "snapshot" fashion, while urine was collected cumulatively. Specifically, it is possible that blood concentrations were increased at some time other than our sampling points, and/or it is possible that because urine formation is essentially a "concentration" process, one can systemically absorb chromium at a rate that is too low to significantly increase blood levels but is sufficient to enhance urinary chromium levels. Also, the precision and accuracy of the analytical determination of total chromium in plasma and RBCs may limit the ability to distinguish small magnitude changes in chromium content at the lower dosages (Anderson *et al.*, 1993). Regardless, these data suggest that at exposures to water concentrations up to 5 mg Cr(VI)/l (50 times the MCL), evaluation of a cumulative urine sample (all voids collected over a period of 24 h or more) will likely be a more sensitive marker for exposure than a single blood sample.

The third and most significant finding of this study is that the RBC and plasma chromium profiles are consistent with systemic absorption of Cr(III), not Cr(VI). Specifically, at the highest dose RBC chromium levels decreased rapidly on cessation of dosing, with an approximate 50% decrease occurring within 7 d post-dosing. It has been hypothesized that such a decline is evidence of systemic absorption of Cr(III) and not Cr(VI) (e.g., Coogan *et al.*, 1991). The persistent elevation in RBC chromium concentrations 14 d following the 10,000 μg-Cr(VI)/d dose (Subjects #1 and #4) might seem to suggest Cr(VI) binding in RBCs. However, when considered in conjunction with the plasma chromium concentrations, which parallel the RBC chromium concentrations, it is more likely that the sustained RBC chromium levels are due to an equilibration of Cr(III) with plasma, blood cell, and tissue compartments. Specifically, if the elevated RBC concentrations at these post-dosing days was due to covalently bound Cr(VI), then one would not expect the "bound" RBC chromium and "free" plasma chromium (i.e., chromium that is being cleared by the kidney) concentrations to follow an identical time-concentration profile. Rather, one would expect the "bound" RBC chromium concentrations to remain at a slowly declining plateau, while the "free" chromium plasma levels decreased to background, as has been shown to occur in animal studies where Cr(VI) was administered intravenously or intratracheally (e.g., Weber, 1983; Edel and Sabbioni, 1985; Gray and Sterling, 1950).

The rate of elimination of chromium from the plasma compartment is also consistent with systemic uptake of Cr(III). Aitio *et al.* (1988) suggested that the half-life of Cr(III) is based on distribution and elimination rates of chromium from three separate compartments: the "fast" elimination compartment ($t_{1/2} = 7$ h), the "moderate" elimination compartment ($t_{1/2} = 15$ d), and the "slow" elimination compartment ($t_{1/2} = 3$ years) (Aitio *et al.*, 1988). Lim *et al.* (1983) reported a similar compartment model, suggesting Cr(III) half-lives of 0.5 to 12 h in blood (fast compartment), 1 to 14 d in storage organs such as the liver and spleen (moderate), and 3 to 12 months in other solid tissues (slow compartment) (Lim *et al.*, 1983). The plasma data for all five subjects at doses of up to 1000 μg-Cr(VI)/d would seem to be consistent with distribution of Cr(III) only to the blood compartment, with rapid urinary excretion of chromium and return to background chromium levels in blood within three or four half-lives (e.g., 24 to 42 h). At the higher doses (5000 and 10,000 μg-Cr(VI)/d), it is possible that Subject #1 and possibly #4 absorbed sufficient Cr(III) such that distribution to a secondary storage compartment occurred. However, it is acknowledged that the data collected in this study do not permit a rigorous analysis of the kinetics of chromium uptake and distribution in the subjects.

For environmental risk assessment purposes, it is important to note that Cr(VI) in water begins to develop a yellow color distinguishable at concentrations of 0.5 to 2.0 mg Cr(VI)/l and becomes bright yellow (and therefore aesthetically objectionable) at concentrations of 2.0 mg Cr(VI)/l. Therefore, chronic exposures to drinking water containing the higher concentrations tested in these experiments would seem unlikely. If the hypothesis regarding gastric reduction of Cr(VI) concentrations is true, then it is possible that Cr(VI) in groundwater may not pose a serious health threat in most cases, even if the MCL is exceeded by 10- to 20-fold. Of course, there are numerous variables in human behavior patterns that would need to be examined to strengthen this assertion. These include the effect of dietary status during Cr(VI) consumption (e.g., composition of diet, fasting vs. non-fasting), ingestion of volumes greater than 1 l/d or greater than 333 ml per dose, and individuals who may be on medication that affects the reductive capacity of the stomach (e.g., antacids). The lack of consistent blood sampling data points throughout each dose for each subject has limited the interpretation of the data, particularly with respect to detailed interindividual comparisons and rigorous statistical analyses. Still, the data that were ultimately collected yielded qualitative and quantitative patterns that strongly support the conclusions given here.

None of the five subjects experienced any adverse health effects as a result of ingesting the designated dosages of Cr(VI), and the three individuals ingesting the higher doses showed no clinically significant changes in urine, blood, or blood chemistry parameters. The absence of clinical findings in this study is consistent with studies cited by the USEPA in support of their current health advisories and regulatory guidelines concerning Cr(VI) (IRIS, 1995).

CONCLUSIONS

The results of this study are consistent with the hypothesis that humans ingesting Cr(VI), at concentrations up to 10.0 mg Cr(VI)/l, absorb Cr(III) systemically because of gastric reduction of the ingested Cr(VI). Hence, consumption of the Cr(VI) drinking water standard of 0.10 mg Cr(VI)/l would also be associated with systemic uptake of Cr(III) and not Cr(VI), and this supports USEPA's assertion that the federal drinking water standard of 0.10 mg Cr(VI)/l is "safe".

CASE STUDY #5: Systemic Uptake of Chromium in Human Volunteers Following Dermal Contact with Hexavalent Chromium: Implications for Risk Assessment

BACKGROUND

Just as chromium speciation is an important factor in understanding system absorption following oral uptake (Case Study #4), the valence state of chromium governs the rate of dermal uptake into blood and tissues (Spruit and van Neer, 1966; Wahlberg, 1970; Hostynek *et al.*, 1993). The Cr(VI) species is considerably more reactive than most Cr(III) species and is also known to be more readily transported across cell membranes when compared with most Cr(III) species and other ionic heavy metals (Wahlberg and Skog, 1965; Spruit and van Neer, 1966; Samitz *et al.*, 1967; Hostynek *et al.*, 1993). Therefore, the valence state of chromium dictates the rate of dermal absorption and systemic uptake. However, substantial quantities of organic reducing agents in the skin matrix can readily convert Cr(VI) to Cr(III). Because most Cr(III) compounds are far less soluble and permeable, the concentration-dependent reduction of Cr(VI) can greatly attenuate penetration of the metal to the deeper skin layers where vascular uptake occurs (Samitz and Katz, 1964; Gammalgaard *et al.*, 1992).

Several investigators have examined the rate of dermal penetration of Cr(VI) compounds at relatively high concentrations both *in vivo* and *in vitro*; however, none of these studies specifically characterized the kinetics of Cr(VI) uptake into the blood. As described in Case Study #4, because trivalent and hexavalent chromium compounds exhibit different patterns of distribution following uptake into human blood (Lewalter *et al.*, 1985; Coogan *et al.*, 1991), the valence state of the absorbed chromium can be distinguished by using RBC chromium concentrations over time as a specific biomarker for Cr(VI) uptake.

The USEPA currently recommends use of a dermal uptake estimate of 1 μg-Cr(VI)/cm^2-h, derived from Baranowska-Dutkiewicz (1981), for risk assessment purposes (Baranowska-Dutkiewicz, 1981; USEPA, 1992). However, the Baranowska-Dutkiewicz (1981) study suffers from design flaws that may grossly overestimate the plausible uptake of Cr(VI) beyond the skin. Specifically, that study measured the amount of chromium recovered from small volumes of Cr(VI) solutions (500 mg Cr(VI)/l) applied to human forearm skin under a watch glass for short time periods. The use of small skin surface areas and an unverified percent recovery in this study seem to create a high degree of uncertainty regarding the estimated dermal absorption rate. Further, Baranowska-Dutkiewicz did not distinguish how much of the Cr(VI) was immediately reduced after contact with the skin, or was it determined whether penetration of Cr(VI) to vascular layers had occurred (i.e., systemic dose).

The purpose of this study was to quantitatively evaluate the systemic uptake of chromium following prolonged dermal contact with water containing an elevated concentration of Cr(VI). For the purposes of this study, a target concentration of 20 mg Cr(VI)/l was used. At water concentrations of 10 mg Cr(VI)/l and greater, the water is obviously bright yellow, and therefore use of a water concentration of 20 mg Cr(VI)/l represents a plausible "worst-case" scenario under which extended contact with tapwater might occur. The goal of this study was to determine whether appreciable uptake of Cr(VI) is likely to occur following a 3-h soaking event and to provide data and insights regarding the dermal exposure assessment of Cr(VI) for risk assessment purposes. Additional detail regarding the design and conduct of this study may be found in Corbett *et al.* (1997).

Chromium uptake and elimination were evaluated in four human adult male volunteers (ages 20 to 44 years) following patch testing to determine if the volunteers had been sensitized previously to chromium. None of the volunteers exhibited a reaction to the patch test or reported adverse health effects from exposure to Cr(VI) in water.

Potassium dichromate, $K_2Cr_2O_7$, was added to chlorinated water in a heated bath (approximately 450 gallons). The final concentration was 22 mg Cr(VI)/l. The free chlorine content of the water was 100 ppm, the initial water temperature was 86°F, and a recirculating heater maintained an average water temperature of 91 ± 2°F based on subsequent readings every 30 min. Samples were collected for analysis of Cr(VI) prior to adding the $K_2Cr_2O_7$, and at time zero, 1, 2 h, and at the end of the exposure (3 h). The concentration of Cr(VI) in the bath water remained relatively constant at 22 ± 1 mg Cr(VI)/l for the duration of the bathing exposure (3 h). Each participant entered the tub at time zero and remained submersed below the shoulders for 3 full hours. Participants showered 45 min after the exposure event.

The methods described in Case Study #4 were utilized to sample and quantify total chromium in blood and urine. Urine voids were collected starting from day 1 (beginning with the first morning void) through the last day of the study (including the last void before bedtime). All urine samples were assayed separately for total chromium by USEPA Method 218.2. Blood samples were collected by venipuncture into Becton-Dickinson Royal Blue Top vacutainer tubes containing EDTA (guaranteed low metals content; <0.4 µg/l) on days 0 (the day prior to exposure), 1, 2, and 4 of the study. All plasma and red blood cell samples were analyzed using a modification of USEPA Method 218.2.

The dermal uptake rate (µg-Cr/cm²-skin-h) was defined as the total amount of chromium excreted above daily historical background levels for each volunteer (taken as the sum of the µg-Cr/d excreted in urine for each volunteer less his historical background excretion) divided by the estimated surface area exposed for each volunteer (cm²) and the duration of the exposure (3 h).

$$\text{Dermal Uptake Rate} = \frac{\text{Total Urinary Chromium Output}}{\text{Surface Area Exposed} * \text{Exposure Period}}$$

The estimates of the maximum dermal uptake rate for systemic absorption of chromium were based on the assumption that any chromium in urine above background levels during days 1 through 5 was related to dermal uptake due to exposure during the study. The total surface area for each volunteer was calculated using a reference nomogram (ICRP, 1984). It was assumed that the portion of the body encompassing 3/4 of the trunk (28% of the total body surface area) and all of the lower limbs (36% of the total body surface area) was exposed for the duration of the 3-h study for each volunteer.

A comparison value for dermal Cr(VI) uptake was also calculated using the USEPA guidance for dermal exposure assessment (USEPA, 1992). For dermal exposures to inorganic substances in aqueous media, the USEPA assumes that the dose per event (DA_{event}) is proportional to the concentration of the chemical in the aqueous media (C_w), the penetration rate of the chemical in skin (k_p), and the length of the exposure (t_{event}). Currently, the USEPA recommends the use of an estimated dermal uptake rate of 1 μg-Cr(VI)/cm^2-h (equivalent to a k_p of 2.1×10^{-3} cm/h). The total dose absorbed (DA_{total}) per event is the product of the dose per event and the surface area to which the chemical was applied.

RESULTS

Urinary Excretion

The average and range of background urinary chromium concentrations for each volunteer are shown in Table 10. Historically, urinary chromium measurements exhibited a moderate degree of variability, likely related to dietary influences. A few high outlier values tended to increase the overall variability.

Total chromium concentrations in the urine for each volunteer are shown in Figure 3. When expressed as total daily excretion of chromium in the urine (μg/d for each person), it appears that all four volunteers may have excreted a small incremental amount of chromium within 1 to 2 d after the bathing event. Volunteer H1 excreted the greatest amount of chromium in the urine following exposure compared with the other volunteers, which is consistent with his higher background excretion rate. In contrast, the other three volunteers eliminated only slightly more chromium than what was typical based on historical background samples.

Chromium Concentrations in Red Blood Cells and Plasma

The red blood cell and plasma concentrations of total chromium for each volunteer are shown in Figure 4. Transient increases in total chromium concentrations in both RBCs and plasma were observed during and immediately following the exposure for all volunteers. The increase in plasma chromium concentrations was generally parallel to that for RBCs. Volunteer H1 exhibited higher chromium concentrations in RBCs and plasma compared with the other volunteers, consistent with his greater chromium excretion in urine.

Dermal Uptake Rate

Table 10 provides a description of the exposed surface areas and the estimated maximum plausible dermal uptake rate for systemic absorption of chromium for

TABLE 10
Estimated Systemic Uptake Rate of Total Chromium Following 3 Hours of Dermal Exposure to 22 mg Cr(VI)/l in Humans

Subject no.	Exposed surface area[a] (cm²)	Historical background[b] (μg/d) mean (range)	Estimated systemic uptake rate at 22 ppm[e] (μg/cm²-h)	Total chromium excretion above historical background[c] (μg)	Total dose absorbed using USEPA Method[d] (μg)
H1	14,336	1.5 (0.25 – 2.6)	4.1E–04	17.5	1,987
H4	13,952	0.5 (0.3 – 1.4)	3.3E–05	1.4	1,934
H5	13,312	0.6 (0.2 – 0.93)	7.5E–05	3.0	1,845
H7	11,392	1.0 (0.25 – 1.7)	7.0E–05	2.4	1,579
Average for all subjects	13,383	0.9 (0.25 – 2.6)	1.5E–04	6.1	1,855

[a] Exposed surface area = 64% of total body surface area, (legs = 36%, 3/4 trunk = 28%) based on nomogram using weight and height to determine surface area (ICRP, 1984).

[b] Detection limit = 0.5 μg/l.

[c] Total chromium excreted for days 1 to 5 less the average historical background level for each volunteer.

[d] $DA_{event} = C_w \times k_p \times t_{event}$; $DA_{total} = DA_{event} \times$ surface area (USEPA, 1992).

[e] Dermal uptake rate is equal to the 5-day total chromium excreted in urine divided by the surface area exposed and the length of exposure for each volunteer.

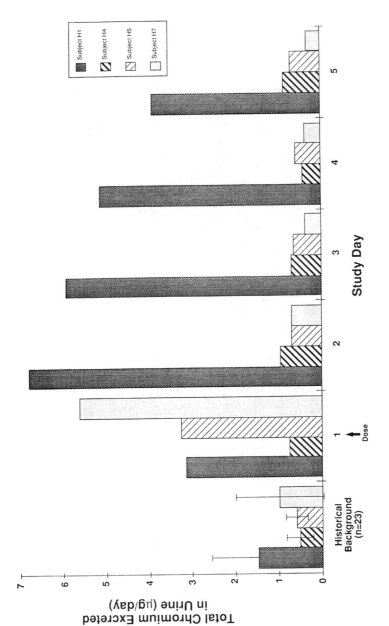

FIGURE 3

Cr concentrations in urine of volunteers following a 3-h dermal exposure to water containing 22 mg Cr(VI)/l. The 3-h dermal exposure occurred between 9:15 a.m. to 12:15 p.m. on day 1. For samples below the detection limit (0.5 μg Cr/l), an assumed concentration of 0.25 μg/l was used to calculate urine Cr. Historical background was calculated for the four participants in the study using available background data (n = 23). The standard deviation of the historical background samples is indicated and labeled with error bars. Volunteer H1's comparatively higher Cr excretion is consistent with his elevated background Cr levels in blood and urine based on past experiments.

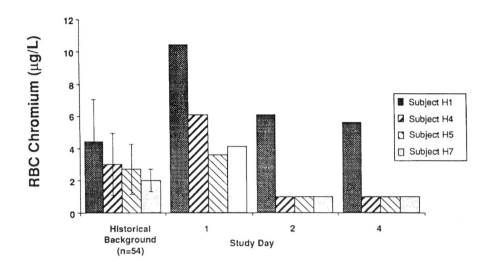

B

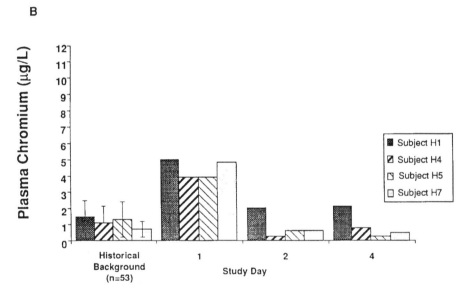

FIGURE 4

Concentrations of total Cr in RBCs and plasma. Total Cr was measured in (A) RBCs and (B) plasma following a 3-h dermal exposure that occurred between 9:15 a.m. and 12:15 p.m. on day 1. For samples below the detection limit (2 µg Cr/l for RBCs and 0.3 to 0.5 µg Cr/ l for plasma) as assumed concentration of half the detection limit was used to calculate Cr levels. Historical background was calculated for the four participants in the study using available background information from other experiments (n = 54 for RBCs; n = 53 for plasma). The standard deviation of the historical background samples is indicated and labeled with error bars. The peak Cr value that occurred during or following exposure was plotted as the day 1 value. Volunteer H1's higher RBC and plasma Cr levels are consistent with his elevated background Cr levels in urine based on past experiments.

695

each volunteer. These systemic uptake rate estimates expressed in μg-Cr/cm^2-h are based on 5-d cumulative urinary chromium excretion adjusted (for historical background) and divided by the volunteer's actual surface area exposed and duration of contact with the water.

The calculated uptake rates based on urinary excretion were all very low, averaging approximately 1.5×10^{-4} $\mu g/cm^2$-h. Volunteer H1 exhibited an apparent dermal uptake rate approximately 7-fold higher than the average for the other three volunteers, which seems to be consistent with the overall trends in urinary and blood measurements of chromium for study days 1 through 5.

For comparison purposes, the total dose absorbed (DA total) was also calculated using the method described in the USEPA guidance for dermal exposure assessment (USEPA, 1992). As shown in Table 10, the total absorbed dose calculated using the USEPA method is approximately 1000-fold greater than the dose measured from urinary excretion of chromium.

DISCUSSION

The current study provides the first quantitative *in vivo* data on systemic uptake of chromium following human exposures in the range of plausible environmental (non-occupational) concentrations of Cr(VI). The data suggest that the apparent rate of systemic uptake of chromium due to dermal contact with water containing 22 mg Cr(VI)/l may be as high as 4.1×10^{-4} μg-Cr/cm^2-h and plausibly as low as zero. In addition, the data show that a 3-h dermal exposure to hexavalent chromium at 22 mg/l does not appear to result in measurable systemic uptake of Cr(VI) based on the premise that RBC chromium concentrations would remain elevated above historical background levels (post-exposure) if appreciable systemic uptake of Cr(VI) had occurred (Lewalter et al., 1985). Both the rate of systemic uptake and the valence state of the systemically absorbed chromium are critical factors in determining plausible health risks from dermal exposure to Cr(VI), especially if the health endpoint of interest involves organs other than the skin.

The dermal uptake rates derived from the current study compare well with the value calculated by Mali et al. (1964) for uptake of 40 mg/l potassium chromate in a cadaver skin diffusion cell system (2×10^{-3} μg-Cr/cm^2-h) (Mali et al., 1964). To date, dermal uptake rates have been derived from artificial test systems using extremely high concentrations of Cr(VI). The dermal uptake rates defined by Wahlberg and Skog (1965) and Mali et al. (1964) were derived from diffusion cell experiments using animal or human skin and prolonged exposures to Cr(VI) (Mali et al., 1964; Wahlberg and Skog, 1965). Samitz and Katz (1964) showed that human skin has a reduction capacity for Cr(VI) of approximately 1 mg/g of skin. It seems likely that the low dermal uptake rate derived here and by Mali et al. (1964) agree because the exposure concentrations of Cr(VI) did not exceed the reductive capacity of the intact skin. At low concentrations, systemic uptake is the key to evaluating the risks, other than dermal effects, that Cr(VI) poses to potentially exposed populations. Therefore, the dermal uptake rate derived here is more appropriate for determining the long-term hazards from plausible drinking water concentrations of Cr(VI).

As noted earlier, the current dermal absorption rate for Cr(VI) recommended by the USEPA contains significant uncertainty. The data obtained in this study indicate that the amount of chromium absorbed into the bloodstream and excreted in the urine is at least two orders of magnitude lower than that calculated using the current USEPA approach. Using the USEPA method (1992) (with the penetration rate derived by Baranowska-Dutkiewicz [1981]), the average dose received by the subjects tested in our study would be 1850 µg, approximately 300-fold higher than the average dose measured from urinary excretion of total chromium in the volunteers (6.1 µg). This discrepancy is attributable to the fact that the current USEPA model does not account for the bioavailability of chromium or the reductive capacity of skin at lower Cr(VI) concentrations that are more plausible for environmental exposure scenarios. Aside from the methodological flaws of the study used to derive the current USEPA absorption rate, the toxicity endpoints that the risk assessment process aims to preclude (liver and kidney toxicity based on high-dose oral toxicity studies) (USEPA, 1996) are only addressed by examining accurate systemic uptake of Cr(VI) and/or Cr(III) compounds.

CONCLUSIONS

The results of this study and those of previous investigators examining relevant environmental concentrations of Cr(VI) in water suggest that the anatomical structure and reducing capacity of the skin provide an effective barrier that prevents any appreciable uptake of Cr(VI).

SUMMARY

The results of the research described in these five case studies have significantly advanced our understanding of how to more accurately assess the human health risks associated with exposure to Cr(VI) in soils and groundwater. Specifically, the following can be concluded:

- Soil concentrations of approximately 1240 ppm Cr(VI) or less will not elicit ACD in a vast majority of the general population (>99.9%), and soil concentrations much higher than this value will also be health protective if the Cr(VI) is not readily bioavailable.

- Soil concentrations up to 400 ppm total chromium are unlikely to influence urinary chromium levels and biomonitoring studies need to consider this probability.

- The human gastrointestinal tract can reduce ingested Cr(VI) to Cr(III) at concentrations up to 10 mg Cr(VI)/l. As water becomes bright yellow at concentrations of 5 mg Cr(VI)/l and greater, oral exposure to Cr(VI) in tapwater is unlikely to pose a significant risk in most situations.

- At water concentrations of up to approximately 22 mg Cr(VI)/l, dermal penetration of Cr(VI) is negligible even under extreme exposure conditions. Therefore, Cr(VI) in tapwater or in standing water in the environment is unlikely to pose a significant systemic health concern.

The ramifications of these findings are significant and are as follows:

- ACD is not an appropriate health endpoint for setting health-based soil standards.

- In many cases, urinary biomonitoring studies are unlikely to be useful in assessing Cr(VI)-related exposures.

- The USEPA's MCL of 0.10 mg Cr(VI)/l appears to contain a large margin of safety.

- Systemic uptake of Cr(VI) following dermal contact with water or soil does not occur to a degree that warrants quantitative evaluation in a risk assessment.

There are other uncertainties in the Cr(VI) risk assessment process that would benefit from additional research. For example, there is currently no USEPA-verified reference concentration for Cr(VI) and, hence, noncancer inhalation hazards for Cr(VI) are not assessed in a standard fashion. Finley *et al.* (1993) suggested an RfC for Cr(VI) particulates and mists based on the work of Steffe and Baetjer (1965) and Lindberg and Hedenstierna (1983) and Malsch *et al.* (1994) suggested a particulate RfC for Cr(VI) based on Glaser *et al.* (1985; 1990). It is also known that the existing inhalation slope factor for Cr(VI) contains a high degree of uncertainty, in part because the epidemiological data on which the slope factor is based (Mancuso, 1975) did not report Cr(VI) concentrations. The "estimated" Cr(VI) concentrations from Mancuso (1975) are based on very conservative estimates of $Cr(VI)/Cr_{total}$ ratios that might have existed in the workplace. Further, because inhalation is often the primary pathway of concern for Cr(VI) in the environment, a better understanding of background levels of Cr(VI) that exist in different parts of the country and their sources would be beneficial. Finally, a better understanding of the relationship between oral Cr(VI) exposure and urinary chromium output would help determine the utility of conducting biomonitoring studies.

All of the results described in these case studies were obtained using human subjects. While the use of human subjects is not without controversy, in each case the studies were conducted in such a way that the participants were at no significant risk. The studies were conducted in accordance with rigorous testing protocols that had received prior approval from a human use committee. This work is consistent with a general trend towards using human subjects in risk assessment research if: 1) the work can clearly be done at no risk to the participants; and 2) human data will avoid significant margins of uncertainty that would inherently be present in animal data. For example, the USEPA is currently conducting a study wherein human volunteers are ingesting prescribed amounts of mine-tailing wastes in order to understand how exposure to this material influences blood lead concentrations. Clearly, information obtained from these types of studies is valuable because the greatest source of uncertainty in risk assessment, animal-to-human extrapolation, is avoided.

It appears that most of the applied risk assessment research being conducted today is funded by the private sector on an ad hoc basis. These types of studies are typically undertaken to address an uncertainty that if left unexamined will require use of a conservative, default assumption in the assessment. It would seem that an organized national research effort would be of benefit, and possibly a consortium of EPA scientists and private industry will eventually establish a research agenda that would prioritize risk assessment uncertainties. Such research items might include:

- Dermal permeability coefficients for volatile organic compounds in water

- Oral bioavailability estimates for metals in soils

- Quantification of kinetic constants in humans for key chemicals

- Toxicokinetic models for all chlorinated volatile organics and persistent organic compounds

In the interim, scientists and regulators should continue to seek opportunities where applied research can be used to reduce uncertainties in risk assessment methods in a timely and cost-effective manner.

REFERENCES

Aaseth, J., Alexander, J., and Norseth, T. 1982. Uptake of ^{51}Cr-chromate by human erythrocytes — a role of glutathione. *Acta. Pharmacol. et Toxicol.* **50,** 310–315.

Ackerman, A. B. 1978. *Histologic Diagnosis of Inflammatory Skin Disease.* pp. 233–236. Philadelphia, Lea and Febiger.

Adams, R. M., Ed. 1990. ACD. In: *Occupational Skin Disease.* 2nd ed. pp. 26–31. Philadelphia, WB Saunders.

Aitio, A., Javisalo, J., Kiilunen, M., Kalliomaki, P. L., and Kalliomaki, K. 1988. Chromium. In: *Biological Monitoring of Toxic Metals*. pp. 369–382. (Clarkson, T. W., Friberg, L., Nordberg, G. F., and Sager, D. R., Eds.) New York, Plenum Press.

Anderson, F. E. 1960. Cement and oil dermatitis: the part played by chrome sensitivity. *Br. J. Dermatol.* **72,** 108–117.

Anderson, R. A., Polansky, M. M., Bryden, N. A., Roginski, E. E., Patterson, K.Y., Veillon, C., and Glinsmann, W. 1982. Urinary chromium excretion of human subjects: effects of chromium supplementation and glucose loading. *Am. J. Clin. Nutr.* **36,** 1184–1193.

Anderson, R. A., Colton, T., Doull, J., Marks, J. G., Smith, R. G., Bruce, G. M., Finley, B. L., and Paustenbach, D. J. 1993. Designing a biological monitoring program to assess community exposure to chromium: conclusions of an expert panel. *J. Toxicol. Environ. Health* **40,** 555–583.

Armitage, P. and Berry, G. 1987. *Statistical Methods in Medical Research*. 2nd ed. pp. 485–486; 495–496. Oxford, Blackwell Scientific.

ASTM. 1992. American Society for Testing and Materials. *Standard Practice for Reducing Samples of Aggregate to Testing Size. Annual Book of ASTM Standards*. Volume 04.02. Philadelphia, PA.

Avellini, B. A., Kamon, E., and Krajewski, J.T. 1980. Physiological responses of physically fit men and women to acclimation of humid heat. *J. Appl. Physiol.* **49,** 254–261.

Bagdon, R. E. 1989. Dermal absorption to selected chemicals under experimental and human exposure conditions to facilitate risk assessment and the development of standards for soil. I. Chromium. Prepared for the Office of Science and Research, New Jersey Department of Environmental Protection.

Bagdon, R. E. and Hazen, R. E. 1991. Skin permeation and cutaneous hypersensitivity as a basis for making risk assessments of chromium as a soil contaminant. *Environ. Health Perspect.* **92,** 111–119.

Baranowska-Dutkiewicz, B. 1981. Absorption of hexavalent chromium by skin in man. *Arch. Toxicol.* **47,** 47–50.

Bonde, J. P. and Christensen, J. M. 1991. Chromium in biological samples from low-level exposed stainless steel and mild steel welders. *Arch. Environ. Health* **46(4),** 225–229.

Bruynzeel, D. P., van Ketel, W. G., von Blomberg, M., and Scheper, R. J. 1983. Angry back or the excited skin syndrome. *J. Am. Acad. Dermatol.* **8,** 392–397.

Bukowski, J. A., Goldstein, M. D., Korn, L. R., and Johnson, B. B. 1991. Biological markers in chromium exposure assessment: confounding variables. *Arch. of Environ. Health* **46(4),** 230–235.

Bukowski, J. A., Goldstein, M. D., Korn, L. R., Rudakewych, M., Shepperly, D., Gates, D., and McLinden, M. 1992. Chromium exposure assessment of outdoor workers in Hudson County, NJ. *Sci. Total Environ.* **122,** 291–300.

Buono, M. J., McKenzie, B. K., and Kasch, F. W. 1991. Effects of aging and physical training on the peripheral sweat production of the human endocrine sweat gland. *Age Ageing* **20,** 439–441.

Burrows, D. and Calnan, C. D. 1965. Cement dermatitis. II. Clinical aspects. *Trans. St. John Hospital Dermatol. Soc.* **51,** 27–39.

Burrows, D., Ed. 1983. Adverse chromate reactions on the skin. In: *Chromium Metabolism and Toxicity*. pp. 138–158. Boca Raton, FL, CRC Press.

Burrows, D. and Adams, R. M. 1990. Metals. In: *Occupational Skin Disease*, 2nd ed. pp. 349–386. (Adams, R. M., Ed.) Philadelphia, WB Saunders.

Calabrese, E. J., Barnes, R., and Stanek, E. J. 1989. How much soil do young children ingest: an epidemiological study. *Regul. Toxicol. Pharmacol.* **10,** 123–137.

Calabrese, E. J., and Stanek, E. J., III. 1991. A guide to interpreting soil ingestion studies. II. Qualitative and quantitative evidence of soil ingestion. *Regul. Toxicol. Pharmacol.* **13,** 278–292.

Calabrese, E. and Stanek, E. 1995. Resolving intertracer inconsistencies in soil ingestion estimation. *Environ. Health Perspect.* **103,** 454–457.

Carter, V. E., Ed. 1982. *Corrosion Testing for Metal Finishing*. pp. 117–118. Philadelphia, PA, Butterworth Scientific.

Coogan, T., Squibb, K. S., Motz, J., Kinney, P. L., and Costa, M. 1991. Distribution of chromium within cells of the blood. *Toxicol. Appl. Pharmacol.* **108(1)**, 157–166.

Corbett, G., Finley, B., Paustenbach, D., and Kerger, B. 1997. Systemic uptake of chomium in human volunteers following dermal contact with hexavalent chromium (22 mg/l). *J. Exp. Anal. Environ. Epidemiol.* **7(2)**, 179–189.

Davis, S., Waller, P., Buschom, R., Ballou, J., and White, P. 1990. Quantitative estimates of soil ingestion in normal children between the ages of 2 and 7 years: population-based estimates using aluminum, silicon, and titanium as soil trace elements. *Arch. Environ. Health* **45**, 112-122.

De Flora, S., Badolati, D., Serra, G. S., Picciotto, A., Magnolia, M. R., and Savarino, V. 1987. Circadian reduction of chromium in the gastric environment. *Mutat Res.* **192(3)**, 169–174.

De Flora, S. and Wetterhahn, K. 1989. Mechanisms of chromium metabolism and genotoxicity. *Life Chem. Rep.* 7, 169–244.

Donaldson, R. M., Jr. and Barreras, R. F. 1966. Intestinal absorption of trace quantities of chromium. *J. Lab. Clin. Med.* 68, 484–493.

Drinkwater, B. L. and Horvath, S. M. 1979. Heat tolerance and aging. *Med. Sci. Sports* **11(1)**, 49-55.

Duggan, M. J., Inskip, M. J., Rundle, S. A., and Moorcroft, J. S. 1985. Lead in playground dust and on the hands of children. *Sci. Total Environ.* **44**, 65–79.

Edel, J. and Sabbioni, E. 1985. Pathways of Cr(III) and Cr(VI) in the rat after intratracheal administration. *Human Toxicol.* **4**, 409–416.

ESE. 1989. Environmental Science and Engineering. Remedial Investigation for Chromium Sites in Hudson County, New Jersey. Prepared for the State of New Jersey Department of Environmental Protection (NJDEP), Trenton, NJ.

Fagliano, J. and Sarvin, J. E. 1994. *Chromium Medical Surveillance Project, Final Technical Report.* NJDOH, New Jersey Department of Health.

Finley, B., Fehling, K., Falerios, M. and Paustenbach, D. 1993. Field validation for sampling and analysis of airborne hexavalent chromium. *Appl. Occup. Environ. Hyg.* **8(3)**, 191–200.

Finley, B. L., Scott, P. K., and Mayhall, D. A. 1994. Development of a standard soil-to-skin adherence probability density function for use in Monte Carlo analyses of dermal exposure. *Risk Anal.* **14**, 555–569.

Finley, B. L., Scott, P. K., Norton, R. L., Gargas, M. L., and Paustenbach, D. J. 1996. Urinary chromium concentrations in humans following ingestion of safe doses of hexavalent and trivalent chromium: implications for biomonitoring. *J. Toxicol. Environ. Health* **48**, 101–121.

Finley, B. L., Kerger, B. D., Katona, M. W., Gargas, M. L., Corbett, G. C., and Paustenbach, D. J. 1997. Human ingestion of chromium (VI) in drinking water: pharmacokinetics following repeated exposure. *Toxicol. Appl. Pharmacol.* **142**, 151–159.

Fischer, T. I. and Maibach, H. I. 1984. Amount of nickel applied with a standard patch test. *Contact Dermatitis* **11**, 285–287.

Fischer, T. I. and Maibach, H. I. 1985. The thin layer rapid use epicutaneous test (TRUE-Test™), a new patch test method with high accuracy. *Br. J. Dermatol.* **112**, 63–68.

Fischer, T. I. and Maibach, H. I. 1989. Easier patch testing with TRUE-Test™. *J. Am. Acad. Dermatol.* **20**, 447–453.

Gammalgaard, B., Fullerton, A., Avnstorp, C., and Menne, T. 1992. Permeation of chromium salts through human skin *in vitro. Contact Dermatitis* **27**, 302–310.

Gargas, M. L., Norton, R. L., Harris, M. A., Paustenbach, D. J., and Finley, B. L. 1994a. Urinary excretion of chromium following ingestion of chromite-ore processing residues in humans: implications for biomonitoring. *Risk Anal.* **14(6)**, 1019–1024.

Gargas, M. L., Norton, R. L., Paustenbach, D. J., and Finley, B. L. 1994b. Urinary excretion of chromium by humans following ingestion of chromium picolinate: implications for biomonitoring. *Drug Metab. Dispo.* **22(4)**, 522–529.

Gargas, M. L., Harris, M. A., Paustenbach, D. J., and Finley, B. L. 1996. Response Letter-to-the-Editor regarding urinary excretion of chromium following ingestion of chromite-ore processing residues in humans: implications for biomonitoring. *Risk Anal.,* in press.

Geiser, J. D., Jeanneret, J. P., and Delacretaz, J. 1960. Eczema au ciment et sensibilization au cobalt. *Dermatologica* **121,** 1–7.

GENSTAT Reference Manual. 1992. *GENSTAT 5, Release 2.2.* Oxford, Clarendon Press.

Gibson, T. E. and Shelley, W. B. 1948. Sexual and racial differences in the response of sweat glands to acetylcholine and pilocarpine. *J. Invest. Dermatol.* **11,** 137–142.

Glaser, U., Hochrainer, D., Kloppel, H., and Kuhnen, H. 1985. Low-level chromium (VI) inhalation effects on alveolar machrophages and immune function in Wister rats. *Arch. Toxicol.* **57,** 250–256.

Glaser, U., Hochrainer, D, and Steinhoff, D. 1990. Investigation of irritating properties of inhaled Cr(VI) with possible influence on its carcinogenic action. *Environ. Hygiene* **1,** 111–116.

Gray, S. J. and Sterling, K. 1950. The tagging of red cells and plasma proteins with radioactive chromium. *J. Clin. Invest.* **29,** 1604–1613.

Herrmann, F., Prose, P. H., and Sulzberger, W. B. 1952. Studies on sweating. V. Studies of quantity and distribution of thermogenic sweat delivery to the skin. *J. Invest. Dermatol.* **18,** 71–86.

Horowitz, S. B. and Finley, B. L. 1993. Using human sweat to extract chromium from chromite ore processing residue: applications to setting health-based cleanup levels. *J. Toxicol. Environ. Health* **40,** 585–599.

Horowitz, S. B. and Finley, B. L. 1994. Setting health protective soil concentrations for dermal contact allergens: a proposed methodology. *Regul. Toxicol. Pharm.* **19,** 31–47.

Hostynek, J. J. and Maibach, H. I. 1988. Chromium in U.S. household bleach. *Contact Dermatitis* **18,** 206–209.

IARC. 1990. International Agency for Research on Cancer. Chromium, *Nickel and Welding In: IARC Monographs on the Evaluation of the Carcinogenic Risk of Chemicals to Humans.* World Health Organization, Geneva, Switzerland.

ICDRG European Standard Series. 1994. *Contact Dermatitis* **11,** 63–64.

ICRP. 1984. International Commission on Radiological Protection. *Report of the Task Group on Reference Man.* ICRP Publication 23. New York, NY. Oxford, Pergamon Press.

IRIS. 1995. Integrated Risk Information System. *Chromium.* Office of Health and Environmental Assessment, U.S. Environmental Protection Agency, Washington, D. C. Down-loaded from National Library of Medicine on-line service, July 27.

Jackson, E. M. and Goldmen, R., Eds. 1990. *Irritant Contact Dermatitis.* New York, Marcel Dekker.

Jacobs, D. S., DeMott, W. R., Finley, P. R., Horvat, R. T., Kasten, B. L., and Tilzer, L. L. 1994. *Laboratory Test Handbook.* 3rd ed., pp. 522–523; 538–539. Cleveland, OH, Lexi-Comp, Inc.

Kimbrough, R., Falk, H., Stehr, P., and Fries, G. 1984. Health implications of 2,3,7,8-tetrachlorodibenzo-*p*-dioxin (TCDD) contamination of residential soil. *J. Toxicol. Environ. Health* **14,** 47–93.

Kissel, J. C., Richter, K. Y., and Fenske, R. A. 1996. Field measurement of dermal soil loading attributable to various activities: implications for exposure assessment. *Risk Anal.* **16,** 115–125.

Kuno, Y. 1934. *The Physiology of Human Perspiration.* London, Churchill.

Lammintausta, K. H. and Maibach, H. F. 1990. Contact dermatitis due to irritation. In: *Occupational Skin Disease.* 2nd ed. pp. 1–3. (Adams, R. M., Ed.) Philadelphia, WB Saunders.

Lauwerys, R. R. 1983. *Industrial Chemical Exposure: Guidelines for Biological Monitoring.* ISBN 0–931890–10–1. Davis, CA, Biomedical Publications.

Levin, H. M., Brunner, N. J., and Ratner, H. 1959. Lithographer's dermatitis. *JAMA,* **169,** 566–569.

Lewalter, J., Korallus, U., Harzdorf, C., and Weidemann, H. 1985. Chromium bond detection in isolated erythrocytes: a new principle of biological monitoring of exposure to hexavalent chromium. *Int. Arch. Occup. Environ. Health* **55,** 305–318.

Lim, T. H., Sargent, T., III, and Kusubov, N. 1983. Kinetics of trace element chromium(III) in the body. *Am. J. Physiol.* **244(4),** R445-R454.

Lindberg, E. and Hedenstierna, G. 1983. Chrome plating: symptoms, findings in the upper airways, and effects on lung function. *Arch. Environ. Health* **38(6),** 367–374.

Lindberg, E. and Vesterberg, O. 1989. Urinary excretion of chromium in chromeplaters after discontinued exposure. *Am. J. Ind. Med.* **16,** 485–492.

Maibach, H. I. 1981. The E.S.S. excited skin syndrome (alias the "angry back"). In: *New Trends in Allergy.* pp. 208–22. (Ring, D. and Burg, B., Eds.) Berlin, Springer-Verlag.

Maibach, H. I., Ed. 1987. *Occupational and Industrial Dermatology,* 2nd ed. pp. 28–31; 190–210. Chicago, Year Book Medical.

Mali, J. W. H., van Kooten, W. J., van Neer, F. C J., and Spruit, D. 1964. Quantitative aspects of chromium sensitization. *Acta Derm. Venereol.* **44,** 44–48.

Malsch, P. A., Proctor, D. M., and Finley, B. L. 1994. Estimation of a chromium inhalation reference concentration using the benchmark dose method: a case study. *Regul. Toxicol. Pharmacol.* **20,** 58–82.

Mancuso, T. F. 1975. Consideration of chromium as an industrial carcinogen. (International Conference on Heavy Metals in the Environment. Ontario, Canada. October 27–31). 343–356.

Marks, J. G. and DeLeo, V. A. 1992. *Contact and Occupational Dermatology.* St. Louis, Mosby-Year Book.

Marks, J., Belsito, D., DeLeo, V., Fowler, J., Fransway, A., Maibach, H., Mathias, T., Nethercott, J., Rietschel, R., Rosenthal, L., Sherertz, E., Storrs, F., and Taylor, J. 1995. North American Contact Dermatitis Group standard tray patch test results (1992–1994). *Am. J. Contact Dermatitis* **6(3),** 160–165.

Minoia, C. and Cavalleri, A. 1988. Chromium in urine, serum and red blood cells in the biological monitoring of workers exposed to different chromium valency states. *Sci. Total Environ.* **71,** 323–327.

Morris, G. E. 1958. "Chrome" dermatitis: a study of the chemistry of shoe leather with particular reference to basic chromic sulfate. *Arch. Dermatol.* **78,** 612–618.

Mutti, A., Cavatora, A., Pedroni, C., Borghi, A., Gioroli, C., and Franchini, I. 1979. The role of chromium accumulation in the relationship between airborne and urinary chromium in welders. *Int. Arch. Occup. Environ. Health* **43,** 123–133.

NACDG. 1984. North American contact dermatitis group. preliminary studies of the TRUE-Test™ Patch Test System in the United States. *J. Am. Acad. Dermatol.* **21,** 841–843.

Nethercott, J. R. 1982. Results of routine patch testing of 200 patients in Toronto, Canada. *Contact Dermatitis* **8,** 389–395.

Nethercott, J. R. 1990. Practical problems in the use of patch testing in evaluation of patients with contact dermatitis. In: *Current Problems in Dermatology*, pp. 101–123. (Weston W., Ed.) St. Louis, Mosby-Year Book.

Nethercott, J., Paustenbach, D., Adams, R., Fowler, J., Marks, J., Morton, C., Taylor, J., Horowitz, S., and Finley, B. 1994. A study of chromium induced ACD with 54 volunteers: implications for environmental risk assessment. *Occup. Environ. Med.* **51(6),** 371–380.

NJDEP. 1990. New Jersey Department of Environmental Protection. Derivation of a Risk-Based Chromium Level in Soil Contaminated with Chromite-Ore Processing Residue in Hudson County. Trenton, NJ.

NJDEP. 1992a. New Jersey Department of Environmental Protection. Risk Assessment of the ACD Potential of Hexavalent Chromium in Contaminated Soil — Derivation of an Acceptable Soil Concentration. Draft by Dr. Alan Stern. Division of Science and Research, Trenton, New Jersey. June 9.

NJDEP. 1992b. New Jersey Department of Environmental Protection. Letter to Dave Rabbe. Proficiency Demonstration Protocol for the NJDEP Modified Method 3060/7196A for Hexavalent Chromium Analysis, August 20.

Pandolf, K. D., Caderette, B. S., Sawka, M. N., Young, A. J., Francesconi, R. P., and Gonzales, R. R. 1988. Thermoregulatory responses of middle-aged and young men during dry-heat acclimation. *J. Appl. Physiol.* **65,** 65–71.

Pandolf, K. B. 1991. Aging and heat tolerance at rest or during work. *Environ. Aging Res.* **17,** 189–204.

Paustenbach, D. J. 1989. A survey of health risk assessment. In: *The Risk Assessment of Environ. and Human Health Hazards: A Textbook of Case Studies*. (D. J. Paustenbach, Ed.) New York, Wiley.

Paustenbach, D. J., Rinehart, W. E., and Sheehan, P. J. 1991a. The health hazards posed by chromium-contaminated soils in residential and industrial areas: conclusions of an expert panel. *Regul. Toxicol. Pharmacol.* **13**, 195–222.

Paustenbach, D. J., Sheehan, P. J., Lau, V., and Meyer, D. M. 1991b. An assessment and quantitative uncertainty analysis of the health risks to workers exposed to chromium contaminated soils. *Toxicol. Ind. Health* **7**, 159–196.

Paustenbach, D. J., Sheehan, P. J., Paul, J. M., Wisser, L. M., and Finley, B. L. 1992. Review of the ACD hazard posed by chromium-contaminated soil: identifying a "safe" concentration. *J. Toxicol. Environ. Health* **37**, 177–207.

Paustenbach, D. J. 1995. The practice of health risk assessment in the United States (1975–1995): How the U.S. and other countries can benefit from that experience. *Hum. Ecolog. Risk Assess.* **1(1)**, 29–79.

Pirilä V. 1954. On the role of chromium and other trace elements in cement eczema. *Acta Derm. Venereol.* **34**, 137–143.

Randin, J. P. 1987. Pitting potential of stainless steels in synthetic sweat. *Werkstoffe Korros.* **38**, 175–183.

Rietschel, R. L., Marks, J. G., and Adams, R. M. 1989. Preliminary studies of the TRUE Patch Test system in the United States. *J. Am. Acad. Derm.* **21**, 841–843.

Roels, H. A., Buchet, J. P., and Lauwerys, R. R. 1980. Exposure to lead by the oral and the pulmonary routes of children living in the vicinity of a primary lead smelter. *Environ. Res.* **22**, 81–94.

Rook, A., Wilkinson, D. S., Ebling, F. J., Champion, R. H., and Burton, J. L., Eds. 1986. *Textbook of Dermatology*. pp. 350–450. Oxford, Blackwell Scientific.

Rothman, S. 1954. *Physiology and Biochemistry of the Skin*. Chicago, IL, University of Chicago Press.

Rothman, K. J. 1987. *Modern Epidemiology*. Boston, Little, Brown, and Co.

Samitz, M. H. and Katz, S. 1964. A study of the chemical reactions between chromium and skin. *J. Invest. Dermatol.* **43(7)**, 35–44.

Samitz, M. H., Katz, S., and Shrager, J. D. 1967. Studies of the diffusion of chromium compounds through skin. *J. Invest. Dermatol.* **48**, 514.

Sheehan, P. J., Meyer, D. M., Sauer, M. M., and Paustenbach, D. J. 1991. Assessment of the human health risks posed by exposure to chromium-contaminated soils. *J. Toxicol. Environ. Health* **32**, 161–201.

Skog, E. and Wahlberg, J. E. 1969. Patch testing with potassium dichromate in different vehicles. *Arch. Dermatol.* **99**, 697–700.

Spruit, D. and van Neer, F. C. J. 1966. Penetration rate of Cr(III) and Cr(VI). *Dermatologica* 132, 179–182.

Steffee, C. H. and Baetjer, A. M. 1965. Histopathologic effects of chromate chemicals. *Arch Environ Health* **11**, 66–75.

Stern, A. H., Freeman, N. C. G., Plesan, P., Bossch, R. R., Wainman, T., Howell, T., Shupack, S. I., Johnson, B. B., and Lioy, P. J. 1992. Residential exposure to chromium waste — urine biological monitoring in conjunction with environmental exposure monitoring. *Environ. Res.* **58**, 147–162.

Storrs, F. J., Rosenthal, L. E., Adams, R. M., Clendenning, W., Emmett, E. A., and Fisher, A. A. 1989. Prevalence and relevance of allergic reactions in patients patch tested in North America — 1984 to 1985. *J. Am. Acad. Dermatol.* **20**, 1038–1044.

Sulzberger, M. B. and Herrmann, F. 1954. *Disturbances in the Delivery of Sweat*. (Curtis, A., Ed.) pp. 13–16. Springfield, IL, Charles C. Thomas.

Tassavainen, A., Nurminen, M., Mutanen, P., and Tola, S. 1980. Application of mathematical modelling for assessing the biological half-times of chromium and nickel in field studies. *Br. J. Ind. Med.* **37**, 285–291.

Tola, S., Kilpiö, J., Virtamo, M., and Haapa, K. 1977. Urinary chromium as an indicator of the exposure of welders to chromium. *Scand. J. Work Environ. Health* **3**, 192–202.

USEPA. 1986. U.S. Environmental Protection Agency. *EPA Method 3060.* SW–846. 2nd ed. September. Washington, DC.

USEPA. 1989. U.S. Environmental Protection Agency. *Exposure Factors Handbook.* Office of Remedial Response. EPA/540/1–88/001. Washington, D.C.

USEPA. 1991. U.S. Environmental Protection Agency. *Federal Register,* January 30, 1991. Volume 56, No. 20.

USEPA. 1992. U.S. Environmental Protection Agency. *Dermal Exposure Assessment: Principles and Applications.* Washington, DC, Office of Research and Development.

USEPA. 1996. U.S. Environmental Protection Agency. *Chromium.* Washington, D.C., Integrated Risk Information System (IRIS). Office of Health and Environmental Assessment. Down-loaded from National Library of Medicine on-line service, January 25.

Veillon, C. 1982. Chromium in urine as measured by atomic absorption spectrometry. *Clin. Chem.* **28**, 2309–2311.

Wahlberg, J. E. and Skog, E. 1965. Percutaneous asbsorption of trivalent and hexavalent chromium. *Arch. Dermatol.* **92**, 315–318.

Wahlberg, J. E. 1970. Percutaneous absorption of trivalent and hexavalent chromium (^{51}Cr) through excised human and guinea pig skin. *Dermatologica* **141**, 288–296.

Wainman, T., Hazen, R., and Lioy, P. 1994. The extractability of Cr(VI) from contaminated soil in synthetic sweat. *J. Exp. Anal. Environ. Epidemiol.* **4(2)**, 171–181.

Weber, H. 1983. Long-term study of the distribution of soluble chromate–51 in the rat after a single intratracheal administration. *J. Toxicol. Environ. Health* **11**, 749–764.

Welinder, H., Littorin, M., Gullberg, B., and Skerfving, S. 1983. Elimination of chromium in urine after stainless steel welding. *Scand J. Work Environ. Health* **9**, 397–403.

WHO. 1988. World Health Organization. Environmental Health Criteria 61: Chromium. Geneva, Switzerland.

Wiegand, H. S., Ottenwalder, H., and Bolt, H. M. 1985. Fast uptake kinetics *in vitro* of ^{51}Cr(VI) by red blood cells of man and rat. *Arch. Toxicol.* **57**, 31–34.

Winston, J. R. and Walsh, E. N. 1951. Chromate dermatitis in railroad employees working with diesel locomotives. *JAMA,* **147**, 1133–1134.

Wright, R. W. 1991. Evaluation of contact dermatitis using the TRUE Patch Test. *J. Arkansas Med. Soc.* **88**, 271–272.

Yang, J. J., Roy, T. A., Krueger, A. J., Neil, W., and Mackerer, C. R. 1989. *In vitro* and *in vivo* percutaneous absorption of benzo(*a*)pyrene from petroleum crude-fortified soil in the rat. *Bull. Environ. Contam. Toxicol.* **43**, 207–214.

Yousef, M. K., Dill, D. B., Vitez, T. S., Hillgard, S. D., and Goldman, A. S. 1984. Thermoregulatory responses to desert heat: age, race, and sex. *J. Gerontol.* **39**, 406–414.

Zelger, J. 1964. Zur Klinik Und Pathogenses Dis Chromate Ekzems. *Arch. Clin. Exp. Dermatol.* **218**, 499–542.

Zelger, J. and Wachter, H. 1966. Uber Die Beziehungen Zwischen Chromat-und Dichromat-allergie. *Dermatologica* **132**, 45–50.

Journal of Soil Contamination, 6(6):707–731 (1997)

Evaluation of 10% Minimum Elicitation Threshold for Cr(VI)-Induced Allergic Contact Dermatitis Using Benchmark Dose Methods

P. K. Scott and D. M. Proctor

ChemRisk Division, McLaren/Hart, Inc., Two Northshore Center, Pittsburgh, Pennsylvania 15212

Historical patch test data have been used to propose health-based soil cleanup levels for Cr(VI) that are protective of eliciting allergic contact dermatitis (ACD) among previously sensitized individuals. Shortcomings regarding the use of these historical studies in the risk assessment of Cr(VI) have been identified and include the use of concentration as the dosimetric for ACD elicitation rather than the mass per surface area. Information on the surface areas of the patches used by the authors of three of the historical studies have been made available recently, and their dose levels have been converted from units of concentration to mass per surface area. For this study, benchmark dose methods were used to estimate the 10% minimum elicitation threshold (MET) based on the converted patch test data from these historical stud-

ies and from the data presented in a more recent patch test study by Nethercott et al. (1994). A truncated lognormal model was fitted to the historical data from each individual historical patch test study, and to the data from the Nethercott et al. (1994) study using maximum likelihood methods. The 10% MET from the Nethercott et al. (1994) study is seven times lower than those from the historical studies. There are two primary reasons for this result. First, Nethercott et al. used a 0.25% potassium dichromate patch to screen study participants, whereas the historical studies used patches with up to 0.5%. Hence, individuals who were less senstive and those who had irritant, rather than allergic reactions.at the high doses, were excluded. Second, Nethercott et al. used a TRUE-Test patch that is a more efficient and reliable allergen delivery device than those used in the historical studies. Assuming 100% bioavailability, the 10% MET from Nethercott et al. (1994) produces an ACD-based soil standard of 445 mg/kg compared with the ACD-based soil standards of 2,750 to 62,500 mg/kg calculated using the historical studies. The most recent patch study of Nethercott et al. (1994), which is based on modern patch testing methods and standardized diagnostic criteria, is the most scientifically appropriate for use in the risk assessment of Cr(VI) and produces the most conservative estimate of the 10% MET.

1058-8337/97/$.50

S OIL cleanup standards for hexavalent chromium [Cr(VI)] are traditionally based on the inhalation and ingestion exposure pathways for the protection of chronic toxicity, including lung cancer (USEPA, 1996). Exposure to high concentrations of Cr(VI) in some occupations (e.g., leather tanners, lithographers, wet cement workers) has been shown to produce skin ulcers as well as irritant and allergic contact dermatitis (ACD) in some individuals (ACGIH, 1989; Hunter, 1974; Lammintausta and Maibach, 1990). Cr(VI)-induced ACD is a Type IV cell-mediated immune response (Adams, 1990) similar to ACD from poison oak. In order to ensure that health-based Cr(VI) soil standards protect against elicitation of ACD, the New Jersey Department of Environmental Protection (NJDEP) has reviewed the results of clinical skin-patch test studies in order to derive an "ACD-based" soil standard (NJDEP, 1995).

Several historical patch test studies of Cr(VI) have been published, primarily in the 1960s (Pirila, 1954; Anderson, 1960; Geiser *et al.*, 1960; Zelger, 1964; Burrows and Calnan, 1965; Zelger and Wachter, 1966; Skog and Wahlberg, 1969; Allenby and Goodwin, 1983). Each of these studies involved patch testing groups of Cr(VI)-sensitized individuals with several different doses (concentrations) of Cr(VI). The numerous shortcomings associated with the use of those patch test studies to develop an ACD-based Cr(VI) soil standard have been discussed in detail by Paustenbach *et al.* (1992). Briefly, much of the historical patch test data for Cr(VI) was collected before the improved and standardized diagnostic criteria developed by the North American Contact Dermatitis Group (NACDG) and the International Contact Dermatitis Research Group (ICDRG) (NACDG, 1984; ICDRG, 1994). Further, these historical studies often failed to disclose information on the diagnostic criteria used to determine allergy, the duration of application, and the analytical methods used to verify the chromium concentrations and valence state. Hence, some irritant reactions were probably scored as allergic responses in these studies (Paustenbach *et al.*, 1992). Also, it is known that the methods used to prepare the patches were inconsistent, and the variability of applied Cr(VI) concentration may have been as high as an order of magnitude (Fischer and Maibach, 1985). In the historical studies, a "patch" was often simply a gauze pad or piece of linen immersed into aqueous Cr(VI) solution before application under plaster or cellophane tape (Pirila, 1954; Anderson, 1960). In other studies, a measured amount of Cr(VI)-petroleum mixture or solution was loaded into a test chamber and applied with paste or tape (Zelger, 1964; Zelger and Wachter, 1966; Skog and Wahlberg, 1969; Allenby and Goodwin, 1983).

However, the biggest shortcoming associated with these studies is the fact that patch conditions were reported in terms of ppm rather than mg Cr(VI)/cm^2. As has been discussed elsewhere (Horowitz and Finley, 1994; Felter and Dourson,

1997), patch concentrations reported only in terms of mass of Cr(VI) per volume of solution are of limited usefulness for setting soil standards because the mass loading of Cr(VI) onto the skin [mg Cr(VI)/cm²-skin] is unknown. To address this data gap, Nethercott et $al.$ (1994) recently conducted a patch-test study in which the applied dose [mg Cr(VI)/cm²-skin] was carefully constructed and measured. The Nethercott et $al.$ (1994) study was specifically designed to determine a 10% minimum elicitation threshold (MET) for Cr(VI) on a mass per surface area basis for use in risk assessment. The 10% MET is the dose of Cr(VI) that will elicit ACD in 10% of the sensitized population. This study utilized the standardized diagnostic criteria developed by NACDG and the standard TRUE-Test patches. Because of its advantages for use in the dermal risk assessment of Cr(VI), the MET derived from this study in terms of mass per surface area is the most appropriate for use in developing a Cr(VI) soil standard protective of ACD (Felter and Dourson, 1997).

Recently, information on the surface areas of the patches used in three of the historical studies (Zelger, 1964; Zelger and Wachter, 1966; Skog and Wahlberg, 1969) have been made available. In this analysis, the applied Cr(VI) concentrations reported by these three studies are converted to the appropriate mass per surface area dosimetric (applied dose). Based on these doses, the 10% METs for each of these studies are calculated using benchmark dose methods similar to those used to derive the 10% MET for the Nethercott et $al.$ (1994) study. Benchmark dose methods involve the fitting of dose-response data to a statistical model and estimating the benchmark dose based on that model. A benchmark dose is a statistical lower confidence limit for a dose that produces a predetermined response rate of an adverse effect compared with background (USEPA, 1995). The objective of this paper is to present the methods used to estimate the 10% METs for elicitation of ACD due to Cr(VI) exposure in terms of mass per surface area for the historical studies and compare these 10% METs to that estimated in Nethercott et $al.$ (1994). Conversion of the historical patch test data to terms useful for risk assessment expands the database available from which better risk management decisions can be made.

Summary of Patch Test Studies

Nethercott et $al.$ (1994)

In the Nethercott et $al.$ (1994) study, 102 potentially Cr(VI)-sensitized volunteers were identified and patch tested in three separate rounds. For the first round, each individual was tested using a diagnostic Cr(VI) patch containing a dose of 4.4 μg Cr(VI)/cm² to determine whether these individuals were actually Cr(VI) sensitive. Only 54 of these individuals responded positively to the diagnostic patch and were

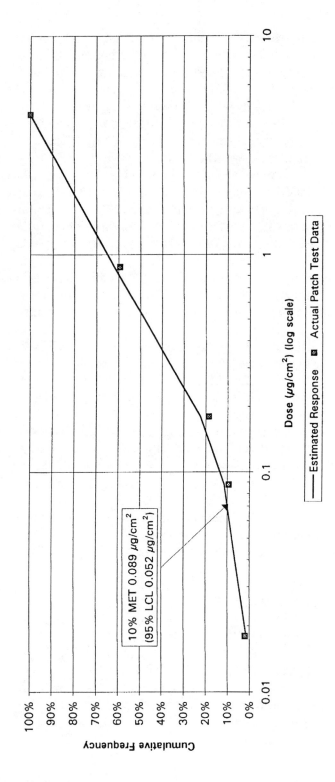

FIGURE 1

Cumulative distribution function for truncated lognormal model. (Data from Nethercott et al. [1994] for potassium dichromate as Cr(VI).)

further patch tested to determine their response level. The other 48 individuals, who were initially identified and thought to be sensitive to Cr(VI) from previous patch tests, were excluded from further analysis. In the second round, the 54 individuals who were positive in Round 1 were patch tested at doses of 0.018 and 0.088 $\mu g/cm^2$. In the third and final round, individuals who had not tested positive to the lower doses used in Round 2 were patch tested at doses of 0.18 and 0.88 $\mu g/cm^2$. All patch testing was performed using the TRUE-Test patch, and responses were scored using the current NACDG diagnostic criteria for patch test interpretation. The responses observed during this study are presented in Table 1. Because this study reported dose in terms of mass per surface area, no concentration-to-dose conversion was necessary. The 10% MET from this study was 0.089 $\mu g/cm^2$. Additional details concerning study design and implementation can be found in Nethercott et al. (1994).

Skog and Wahlberg (1969)

Skog and Wahlberg (1969) tested 46 Cr(VI)-sensitized individuals, aged 23 to 68 years, including 5 women and 41 men. Test patches containing eight concentrations of potassium dichromate ranging in concentration from 10 to 1800 ppm Cr(VI) in three different carrier solutions were applied to the back for 48 h, and the results were recorded 48 h later. The three carrier solutions were water (pH 7), alkaline glycine buffer solution (pH 12), and petrolatum. The authors classified patients according to the lowest test concentration that produced a reaction; however, they did not present their diagnostic criteria for classifying positive reactions.

While the authors did not describe the test patch material used or its method of application in the original paper, recently one of the authors has made available the data needed to convert the Cr(VI) patch concentrations (ppm) to Cr(VI) dose ($\mu g/cm^2$) (Nethercott, personal communication, 1996). For this study, Finn chamber patches, which had a surface area of 0.5 cm^2, were used, and 20 μl of Cr(VI) solution was applied to each patch by micropipette. The following equation describes the conversion from patch concentration (ppm or $\mu g/ml$) to patch dose ($\mu g/cm^2$):

$$\text{Dose} = \frac{[Cr(VI)] * V}{SA} \tag{1}$$

where: Dose is the applied dose of Cr(VI) ($\mu g/cm^2$); [Cr(VI)] is the applied concentration of Cr(VI) ($\mu g/ml$ or ppm); V is the volume of Cr(VI) solution applied to the patch (ml); and SA is the surface area of the patch (cm^2).

TABLE 1

Summary of Nethercott *Et Al.* (1994) Patch Test Study

Chromium compound	Carrier	Cr(VI) dose ($\mu g/cm^2$)	Number responding	Percent responding (%)	Cumulative response	Cumulative percent response (%)
Potassium dichromate	Neutral	0.018	1/54	2	1/54	2
		0.088	4/54	7	5/54	9
		0.18	5/54	9	10/54	19
		0.88	22/54	41	32/54	59
		4.4	22/54	41	54/54	100

TABLE 2
Summary of Skog and Wahlberg (1996) Patch Test Study

Chromium compound	Carrier	Cr(VI) concentration (ppm)	Cr(VI) dose (µg/cm²)	Number responding	Percent responding (%)	Cumulative response	Cumulative percent response (%)
Potassium dichromate	Neutral	10	0.4	1/46	2.2	1/46	2.2
		30	1.2	0/46	0	1/46	2.2
		60	2.4	5/46	11	6/46	13
		100	4.0	8/46	17	14/46	30
		200	8.0	2/46	4.3	16/46	35
		400	16	4/46	8.7	20/46	44
		900	36	6/46	13	26/46	57
		1800	72	20/46	43	46/46	100
Potassium dichromate	Alkaline	10	0.4	3/46	6.5	3/46	6.5
		30	1.2	7/46	15	10/46	22
		60	2.4	8/46	17	18/46	39
		100	4.0	7/46	15	25/46	54
		200	8.0	7/46	15	32/46	70
		400	16	10/46	22	42/46	91
		900	36	1/46	2.1	43/46	94
		1800	72	3/46	6.5	43/46	100
Potassium dichromate	Petroleum	10	0.4	1/46	2.2	1/46	2.2
		30	1.2	2/46	4.3	3/46	6.5
		60	2.4	2/46	4.3	5/46	11
		100	4.0	10/46	22	15/46	33
		200	8.0	10/46	22	25/46	54
		400	16	9/46	20	34/46	74
		900	36	4/46	8.7	38/46	83
		1800	72	8/46	17	46/46	100

The original patch concentrations presented in Skog and Wahlberg (1969), the converted doses, and the responses reported in the study are presented in Table 2. For the purposes of this analysis, dose-response modeling was assessed for each of the three carrier solutions (neutral, alkaline, and petrolatum).

Zelger (1964) and Zelger and Wachter (1966)

Zelger (1964) examined 33 people known to be Cr(VI)-sensitive by patch testing, who ranged in age from 23 to 66 years old. Patients were exposed to 12 different allergens, including potassium dichromate, potassium chromate, chromic acid, and lead dichromate, with Cr(VI) concentrations ranging from 0.5 to 1000 ppm. The potassium dichromate solution was buffered by an acidic carrier solution with pH 1.5, and the potassium chromate, chromic acid, and lead dichromate solutions were buffered by an alkaline carrier solution with pH 11.7. Each patch test appliance, described in the original manuscripts as a Bieiresdorf patch, contained 5 patches, which remained on the back for 24 h, and reactions were read 30 min after removal and again on each of the following 4 d. The reaction threshold was defined as the weakest concentration producing a definite positive reaction, indicated by reddening, swelling, and blistering in the entire contact area.

In the Zelger and Wachter (1966) study, 50 people allergic to Cr(VI), ranging in age from 22 to 72 years, were tested using eight dilutions of potassium chromate and potassium dichromate in an acid-buffered glycine solution of pH 1.5. The concentrations of the patches ranged from 0.5 to 500 ppm Cr(VI). All patients were pretested with 0.5% (5000 ppm) potassium dichromate diagnostic patches to verify their Cr(VI) sensitivity. Testing and diagnostic procedures remained unchanged from the previous Zelger (1964) study except that only potassium dichromate and potassium chromate were tested.

While the patch material used in both studies was not described in the original papers, Dr. Zelger recently has described the type of patch used and the method of application for the Cr(VI) solutions in both studies (Nethercott, personal communication, 1996). The Bieiresdorf patch test appliance used in both studies contained five patches with an approximate surface area of 2 cm^2 per patch. For each patch, 100 µl of Cr(VI) solution was applied by pipette. Using these data with Equation 1, the Cr(VI) concentrations reported in Zelger (1964) and in Zelger and Wachter (1996) were converted into Cr(VI) doses in terms of mass per area. Tables 3 and 4 present the originally reported Cr(VI) concentrations, the converted Cr(VI) doses, and responses for Zelger (1964) and Zelger and Wachter (1966), respectively.

TABLE 3
Summary Of Zelger (1964) Patch Test Study

Chromium compound	Carrier	Cr(VI) concentration (ppm)	Cr(VI) dose (μg/cm²)	Number responding	Percent responding (%)	Cumulative response	Cumulative percent response (%)
Potassium dichromate	Neutral	0.5	0.025	0/33	0	0/33	0
		1.0	0.05	0/33	0	0/33	0
		5.0	0.25	0/33	0	0/33	0
		10	0.5	0/33	0	0/33	0
		100	5.0	2/33	6	2/33	6
		500	25	17/33	52	19/33	58
		1000	50	14/33	42	33/33	100
Potassium chromate	Alkaline	0.5	0.025	0/33	0	0/33	0
		1.0	0.05	0/33	0	0/33	0
		5.0	0.25	0/33	0	0/33	0
		10	0.5	3/33	9.1	3/33	9.1
		50	2.5	9/33	27	12/33	36
		100	5.0	10/33	30	22/33	67
		500	25	11/33	33.3	33/33	100
Chromic acid	Alkaline	0.5	0.025	0/13	0	0/13	0
		1.0	0.05	0/13	0	0/13	0
		5.0	0.25	0/13	0	0/13	0
		10	0.5	1/13	7.7	1/13	7.7
		50	2.5	5/13	38	6/13	46
		100	5.0	2/13	15	8/13	62
		500	25	5/13	38	13/13	100

TABLE 3 (continued)
Summary Of Zelger (1964) Patch Test Study

Chromium compound	Carrier	Cr(VI) concentration (ppm)	Cr(VI) dose (μg/cm²)	Number responding	Percent responding (%)	Cumulative response	Cumulative percent response (%)
Lead chromate	Alkaline	0.5	0.025	0/33	0	0/33	0
		1.0	0.05	0/33	0	0/33	0
		5.0	0.25	0/33	0	0/33	0
		10	0.5	0/33	0	0/33	0
		50	2.5	5/33	15	5/33	15
		100	5.0	8/33	24	13/33	39

TABLE 4
Summary of Zelger and Wachter (1966) Patch Test Study

Chromium compound	Carrier	Cr(VI) concentration (ppm)	Cr(VI) dose (μg/cm²)	Number responding	Percent responding (%)	Cumulative response	Cumulative percent response (%)
Potassium dichromate	Acid	0.5	0.025	0/50	0	0/50	0
		1.0	0.05	0/50	0	0/50	0
		5.0	0.25	0/50	0	0/50	0
		10	0.5	0/50	0	0/50	0
		50	2.5	0/50	0	0/50	0
		100	5.0	2/50	4	2/50	4.0
		500	25	25/50	50	27/50	54
		1,000	50	23/50	46	50/50	100
Potassium chromate	Acid	0.5	0.025	0/50	0	0/50	0
		1.0	0.05	0/50	0	0/50	0
		5.0	0.25	0/50	0	0/50	0
		10	0.5	4/50	8	4/50	8.0
		50	2.5	12/50	24	16/50	32
		100	5.0	16/50	32	32/50	64
		500	25	1/8/50	36	50/33	100

Statistical Methods

To calculate the 10% MET in terms of mass per surface area, a truncated lognormal model was fitted to the patch test data from the four studies evaluated using maximum likelihood methods similar to those used in Nethercott *et al.* (1994) and Paustenbach *et al.* (1992). One dataset, the Zelger (1964) lead dichromate dataset, had too few dose groups with responses (only two) to provide meaningful results, and the model was not fitted to this dataset. Because a reaction at the maximum tested dose was a criterion for inclusion in the patch test study, the distribution of response doses for each study was truncated to the right. In addition, because an individual's actual elicitation threshold will lie between the highest dose for which there was no response and the lowest dose for which there was a response, the maximum likelihood methods appropriate for interval censored data were used (Lawless, 1982). The estimated model parameters were then used to calculate the 10% MET for each dataset. The 95% lower confidence limits (LCLs) of the 10% METs were estimated using the likelihood ratio methods presented by Crump and Howe (1985). The estimated 10% METs and their associated 95% LCLs are presented in Table 5.

The goodness-of-fit of the data to the truncated lognormal model was evaluated using a Chi Square (χ^2) goodness-of-fit test at a 99% confidence level. Two datasets were not tested for goodness-of-fit because the number of dose groups with actual responses (≤ 3) was too low for the application of the test. These two datasets included the Zelger (1964) and the Zelger and Wachter (1966) potassium dichromate datasets. The results of the χ^2 test for the remaining datasets are presented in Table 5.

Results and Discussion

The 10% METs for the historical studies ranged from 0.55 to 12.5 $\mu g/cm^2$, with 95% LCLs ranging from 0.21 to 7.2 $\mu g/cm^2$. Of the datasets that could be evaluated using the χ^2 test, all but one dataset fit the truncated lognormal model at a 99% confidence level, with *p*-values ranging from 0.044 to 0.577. The only dataset that did not fit the model was the Skog and Wahlberg (1969) potassium dichromate in neutral solution dataset, which had a *p*-value of 0.001. All of the historical studies had 10% METs substantially greater than the 10% MET for the Nethercott *et al.* (1994) study of 0.089 $\mu g/cm^2$. Figures 1 through 9 show the actual data and the dose response curves for each of the patch test studies.

The 10% MET for the Nethercott *et al.* (1994) is a factor of seven times lower than the lowest 10% MET from the historical studies. There are two probable explanations as to why the Nethercott *et al.* (1994) study produces such a conser-

TABLE 5

Summary of 10% Minimum Elicitation Thresholds (METs) and Their Associated 95% Lower Confidence Limits (LCLs) for the Patch Test Studies Evaluated

Study[a]	10% MET ($\mu g/cm^2$)[b]	95% LCL of 10% MET ($\mu g/cm^2$)[c]	Chi square p-value	Model fit?[d]
Nethercott et al. (1994)	0.089	0.052	0.713	Yes
Skog and Wahlberg (1969)				
Potassium dichromate, neutral	1.63	0.66	0.001	No
Potassium dichromate, alkaline	0.57	0.34	0.331	Yes
Potassium dichromate, petrolatum	1.4	0.88	0.044	Yes
Zelger and Wachter (1966)				
Potassium dichromate, acid	12.5	7.2	NA	
Potassium chromate, acid	0.72	0.46	0.059	Yes
Zelger (1964)				
Potassium dichromate, acid	10.4	4.7	NA	
Potassium chromate, alkaline	0.63	0.37	0.192	Yes
Chromic acid, alkaline	0.55	0.21	0.577	Yes

a One dataset was excluded due to the low number of dose groups with responses (≤ 2). This dataset included Zelger (1964), lead dichromate.

b 10% MET estimated using maximum likelihood estimates of the parameters of cumulative distribution function for truncated lognormal model.

c 95% LCL of the 10% MET estimated using likelihood ratio method discussed in Crump and Howe (1984).

d The data were considered to fit the lognormal model at a 99% confidence level if the p-value of the Chi square test was greater than 0.01.

vative 10% MET relative to the historical patch test studies. First, and most importantly, the use of a 0.25% potassium dichromate diagnostic patch in the first round of the Nethercott et al. (1994) study vs. the 0.5% potassium dichromate diagnostic patch used in the historical studies may have led to the exclusion of individuals who are less sensitive (have a MET between 0.25 and 0.5%) or excluded those who had an irritant, rather than allergic, response at the higher dose. If either or both are correct, then the 10% MET of the Nethercott et al. (1994) is more conservative and more accurate than those of the historical studies. Indeed, if historical data are analyzed, with exclusion of doses greater than 0.25%, the difference in 10% MET is not significant (e.g., less than 5% for all data sets). This study suggests that the historical data included positive results for individuals who were either not allergic at all (an irritant response) or who were far less sensitive than the patients tested in Nethercott et al. (1994).

Secondly, the TRUE-Test patch used in Nethercott et al. (1994) is a more efficient and reliable allergen delivery device than the Bieiresdorf and Finn chamber patch devices used in the historical studies. The TRUE-Test patches are specifically designed to hydrate by perspiration and are applied to the skin under occlusion that ensures maximal contact with the skin, enabling higher allergen bioavailability (Rietschel et al., 1989). It is possible that older patch test methods (such as the Finn Chamber) did not deliver all of the applied dose to the skin surface, thereby yielding higher estimates of the METs.

As described in Horowitz and Finley (1994), an ACD-based soil standard can be derived as follows:

$$\text{Soil Concentration } (\text{mg - allergen/kg - soil}) = \frac{\text{MET} * \text{CF}}{\text{SA} * \text{BVA}}$$

where: MET is the minimum elicitation threshold determined from applied dose patch-test data (mg-allergen/cm^2-skin); CF is conversion factor (10^6 mg-soil/kg-soil); SA is soil adherence factor (mg-allergen/cm^2-skin); and BVA is bioavailabilibity (unitless).

The USEPA's suggested average value of 0.20 mg soil/cm^2-skin was used to represent the soil adherence factor for this analysis (USEPA). As described in detail in Horowitz and Finley (1994), the 10% MET of 0.089 µg Cr(VI)/cm^2-skin derived from Nethercott et al. (1994) yields an ACD-based soil standard of 445 mg Cr(VI)/kg-soil, assuming 100% bioavailability. The 10% METs identified in the historical studies yield ACD-based soil standards that range from 2750 to 62,500 mg/kg (assuming 100% bioavailability). Therefore, it can be concluded that a soil concentration of 445 ppm Cr(VI) should protect a vast majority of the Cr(VI)-sensitized population from an ACD hazard.

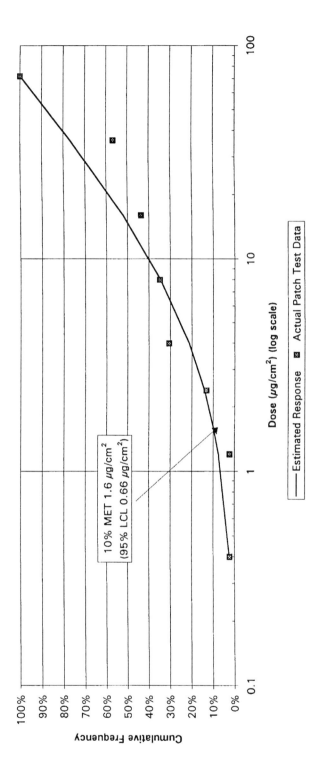

FIGURE 2

Cumulative distribution function for truncated lognormal model. (Data from Skog and Wahlberg [1969] for potassium dichromate as Cr(VI) in neutral solution.)

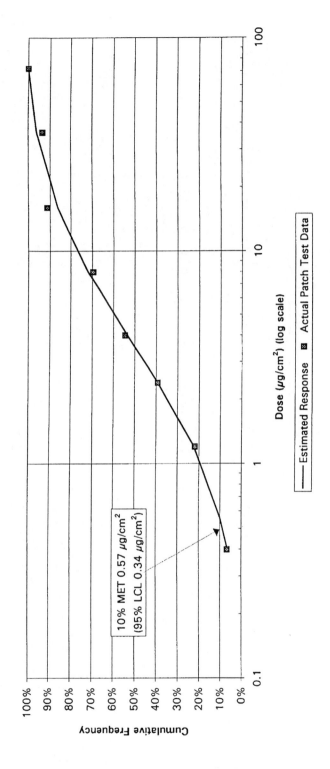

FIGURE 3

Cumulative distribution function for truncated lognormal model. (Data from Skog and Wahlberg [1969] for potassium dichromate as Cr(VI) in alkaline solution.)

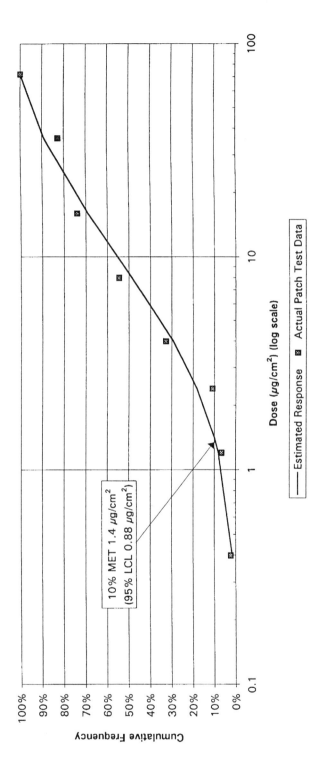

FIGURE 4

Cumulative distribution function for truncated lognormal model. (Data from Skog and Wahlberg [1969] for potassium dichromate as Cr(VI) in petrolatum solution.)

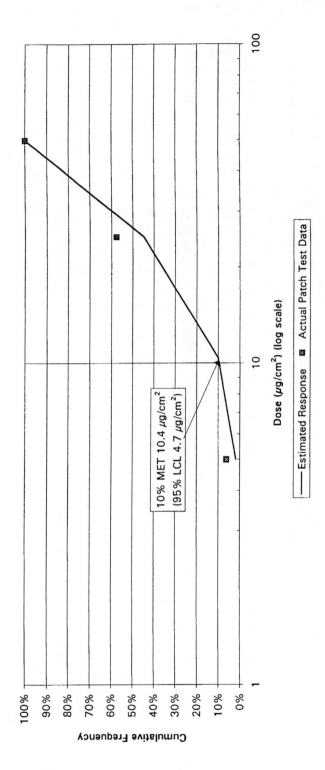

FIGURE 5

Cumulative distribution function for truncated lognormal model. (Data from Zelger [1964] for potassium dichromate as Cr(VI)in acid solution.)

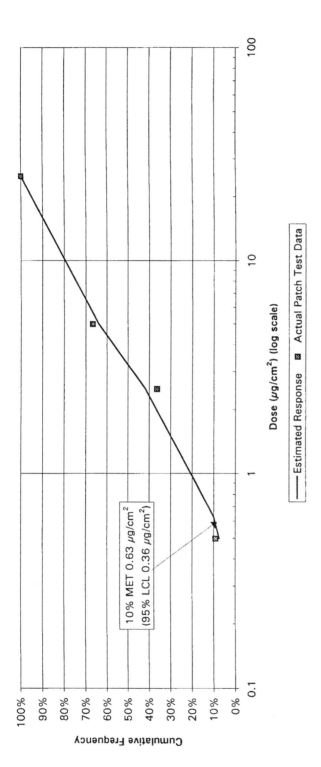

FIGURE 6

Cumulative distribution function for truncated lognormal model. (Data from Zelger [1964] for potassium chromate as Cr(VI) in alkaline solution.)

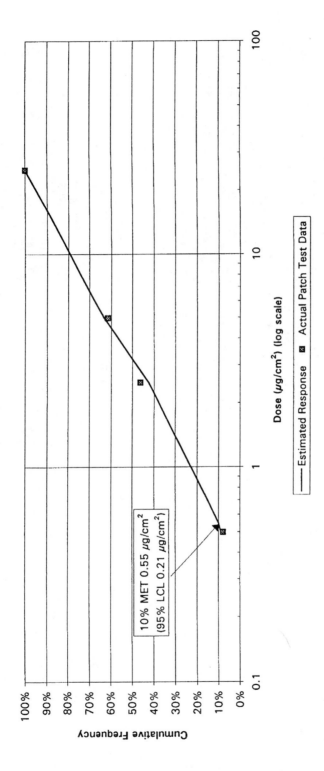

FIGURE 7

Cumulative distribution function for truncated lognormal model. (Data from Zelger [1964] for chromic acid as Cr(VI) in alkaline solution.)

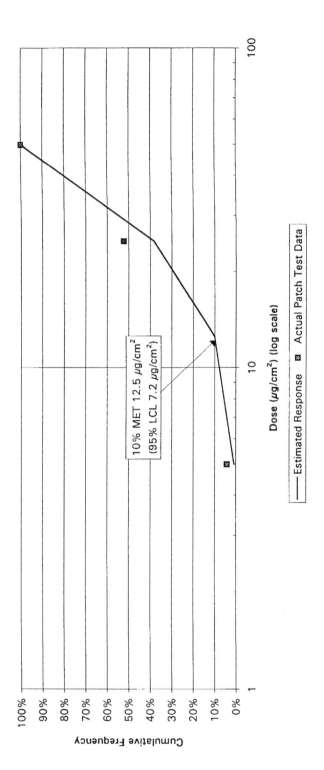

FIGURE 8

Cumulative distribution function for truncated lognormal model. (Data from Zelger and Wachter [1966] for potassium dichromate as Cr(VI) in acid solution.)

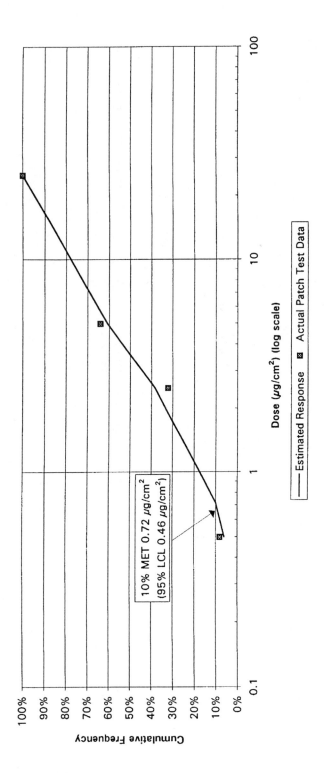

FIGURE 9

Cumulative distribution function for truncated lognormal model. (Data from Zelger and Wachter [1966] for potassium chromate as Cr(VI) in acid solution.)

When the reported Cr(VI) data from historical patch test studies are converted from concentrations (ppm) to mass per surface area ($\mu g/cm^2$-skin), the 10% METs estimated from these older studies are far greater than the estimate obtained from the most recent Nethercott *et al.* (1994) data. Not only is the Nethercott *et al.* (1994) study the most appropriate for use in risk assessment, due to the use of modern patch testing techniques and standardized diagnostic criteria, it also produces the most conservative estimate of 10% MET. Clearly, the 10% MET from the Nethercott *et al.* (1994) study is the most appropriate value for use in assessing the elicitation of ACD in environmental health risk assessment.

Acknowledgment

This manuscript was prepared from information obtained and initially analyzed by Dr. James Nethercott, M.D. These data and analyses were presented, in part, by Dr. Nethercott at the Association for the Environmental Health of Soils (AEHS) Conference (March 12, 1996).

REFERENCES

Adams, R. M. 1990. Job descriptions with their irritants and allergens. In: *Occupational Skin Disease, 2nd ed.*, 578, pp. 666–668. Philadelphia, W.B. Saunders Company.

Allenby, C. F. and Goodwin, B. F. J. 1983. Influence of detergent washing powders on minimal eliciting patch test concentrations of nickel and chromium. *Contact Dermatitis* **9,** 491–499.

American Conference of Governmental Industrial Hygienists (ACGIH). 1989. *Documentation of the Threshold Limits Values and Biological Exposure Indices,* 5th ed. Supplemental Documentation for 1989. Cincinnati, Ohio, ACGIH.

Anderson, F. E. 1960. Cement and oil dermatitis. The part played by chrome sensitivity. *Br. J. Dermatol.* **72,** 108–117.

Burrows, D. and Calnan, C. D. 1965. Cement dermatitis. II. Clinical aspects. *Trans St. John Hosp. Derm. Soc.* **51,** 27–39.

Crump, K. and Howe, R. 1985. A review of methods for calculating statistical confidence limits in low dose extrapolation. In: *Toxicological Risk Assessment* (Vol. 1, *Biological and Statistical Criteria*) (Clayson, D., Krewski, D., and Munroe, I., Eds.) Boca Raton, Florida, CRC Press.

Felter, S. and Dourson, M. 1997. Hexavalent chromium-contaminated soils: Options for risk assessment and risk management. *Regul. Toxicol. Pharmacol.* **25,** 43–59.

Fischer, T. I. and Maibach, H. I. 1985. The thin layer rapid use epicutaneous test (TRUE-test), a new patch test method with high accuracy. *Br. J. Dermatol.* **112,** 63–68.

Geiser, J. D., Jeanneret, J. P., and Delacretaz, J. 1960. Eczema au ciment et sensibilization au cobalt. *Dermatologica* **121,** 1-7.

Horowitz, S. B. and Finley, B. L. 1994. Setting health-protective soil concentrations for dermal contact allergens: a proposed methodology. *Regul. Toxicol. Pharmacol.* **17,** 31–47.

Hunter, D. 1974. *The Disease of Occupations*, 5th ed. Boston, Little, Brown.

ICDRG. 1994. European Standard Series. *Contact Dermatitis.* **11,** 63–64.

Lammintausta, K. H. and Maibach, H. I. 1990. Irritant dermatitis syndrome. *Immunol. Allergic Clin. N. Am.* **9(3),** 435–446.

Lawless, J. F. 1982. *Statistical Models and Methods for Lifetime Data*. New York, John Wiley & Sons.

Nethercott, J., Paustenbach, D., Adams, R., Fowler, J., and Marks, J. 1994. A study of chromium induced allergic contact dermatitis with 54 volunteers: implications for environmental risk assessment. *Occupat. Environ. Med.* **51(6),** 371–380.

Nethercott, J. 1996. Personal Communication. Letter to Deborah Proctor dated March 28, 1996, regarding: water contamination and skin.

New Jersey Department of Environmental Protection 1995. Basis and background for soil cleanup criteria for trivalent and hexavalent chromium. **September.**

North American Contact Dermatitis Group. (NACDG). 1984. North American Contact Dermatitis Group. Preliminary studies of the TRUE-Test™ Patch Test system in the United States. *J. Am. Acad. Dermatol.* **21,** 841–843.

Paustenbach, D. J., Sheehan, P. J., Paull, J. M., Wisser, L. M., and Finley, B. L. 1992. Review of the allergic contact dermatitis hazard posed by chromium-contaminated soil: identifying a "safe" concentration. *J. Toxicol. Environ. Health.* **37,** 177–207.

Pirila, V. 1954. On the role of chrome and other trace elements in cement eczema. *Acta Dermato-Venereol* **34,** 136–143.

Rietschel, R. L., Marks, J. G., and Adams, R. M. 1989. Preliminary studies of the TRUE Patch Test system in the United States. *J. Am. Acad. Dermatol.* **21,** 841–843.

Skog, E. and Wahlberg, J. E. 1969. Patch testing with potassium dichromate in different vehicles. *Arch. Dermatol.* **99,** 697–700.

U.S. Environmental Protection Agency. (USEPA). 1992. *Dermal Exposure Assessment: Principles and Applications.* Washington, D.C. Office of Research and Development.

U.S. Environmental Protection Agency. (USEPA). 1995. *The Use of the Benchmark Dose Approach in Health Risk Assessment.* Washington, D.C. Risk Assessment Forum. EPA/630/R-97/007.

U.S. Environmental Protection Agency. (USEPA). 1996. *Soil Screening Guidance: User's Guide.* Washington, D.C. Office of Solid Waste and Emergency Response. EPA/540/R-96/018.

Zelger, J. 1964. On the pathogenesis and clinical aspects of chromate eczema. *Arch. Klin. Exp. Dermatol.* **218,** 499–502.

Zelger, J. and Wachter, H. 1966. On the relationships between chromate and dichromate allergy. A contribution to the analysis of Chromium(VI) allergy. *Dermatologica* **132,** 45–50.

Journal of Soil Contamination, 6(6):733–749 (1997)

Investigation and Remediation of Chromium and Nitrate Groundwater Contamination: Case Study for an Industrial Facility

Christopher R. Maxwell

Lahontan Regional Water Quality Control Board, 15428 Civic Drive, Suite 100, Victorville, CA 92392

During a preliminary site investigation at an industrial facility (Facility) in the California desert, chromium and nitrate groundwater degradation were identified in the area of a formerly operated landfill. No currently used water supplies were identified as being affected. A groundwater extraction system was installed and operated to contain the plumes and remove contaminant mass. Based on data collected during 5 years of groundwater extraction, it was determined that contaminant reduction was not occurring. A review of disposal records for the Facility revealed that potential sources of chromium and nitrate had not been removed. A detailed field investigation of these areas was subsequently completed. The investigation included trenches, several hundred soil borings, and analysis of soil and groundwater samples. The data collected identified the location and extent of the sources of both the nitrate and chromium groundwater degradation. These sources were previously disposed chromium-bearing kiln bricks in the landfill and undetonated blasting materials in the quarry area at the Facility. The investigation also better defined the geology, hydrology, and contaminant migration behavior. The detailed manner in which the investigation was conducted has provided the basis for the development of a revised corrective action program that coordinates contaminant source removal and groundwater extraction. The focus of this technical article is threefold. First, the article discusses the site investigation that was conducted and the data which were collected at the Facility. The second focus is the final corrective action plan that was developed. Finally, the cost-effective nature of the investigation and cleanup is discussed.

KEY WORDS: *chromium groundwater, nitrate groundwater, plumes, contaminant mass.*

1058-8337/97/$.50

*B*ETWEEN 1953 and 1980, approximately 12,000 cubic yards of solid wastes were discharged to an unlined landfill at an industrial facility. These wastes include several thousand cubic yards of kiln brick. When originally placed in the kiln, the brick contains trivalent chromium [Cr(III)]. In the environment of the kiln some of the Cr(III) on the exterior surfaces of the brick is oxidized to the hexavalent state [Cr(VI)].

In 1986, a preliminary investigation was conducted to evaluate groundwater quality in the area of the landfill. Total chromium and nitrate were detected in groundwater samples, which exceeded both measured background concentrations and California Primary Maximum Contaminant Levels (CMCLs) for drinking water sources. The CMCLs for total chromium and nitrate are 50 µg/l and 45 mg/l, respectively.

Subsequent investigations were conducted between 1987 and 1990 to characterize the hydrogeology in the vicinity of the landfill and define the limits of the total chromium and nitrate groundwater plumes. The plumes of total chromium and nitrate in groundwater, as measured in 1990, are depicted on Figures 1 and 2, respectively. The source(s) of the nitrate plume appear to be at least partially upgradient of the landfill.

In order to promptly contain the plumes of nitrate and total chromium, a groundwater extraction system was designed, installed, and became operational in 1990. The system initially consisted of three extraction wells. Extracted water was conveyed to a holding tank and used as beneficial process water at the Facility. Between 1990 and 1994, the focus of the extraction system was modified to include contaminant mass removal. By 1994, a total of 11 wells were being used for extraction. Maximum extraction rates for single wells ranged from approximately 5 to 7 gallons per minute (gpm). Between 1990 and August 1994, approximately 7.5 million gallons of groundwater were extracted and treated at the Facility.

DRIVING FORCE FOR REMEDIAL ACTIONS

An evaluation of the groundwater usage at the Facility indicated that the impacted groundwater was not currently being used for domestic purposes. No exposure pathways were noted whereby biological receptors, including man, would be exposed to the contaminants. Thus, human health and ecologic risks were not the driving force for remedial actions.

The Water Quality Control Plan for the Lahontan Regional Water Quality Control Board (Basin Plan) prescribes beneficial uses for groundwaters and contains narrative objectives for protection of these uses. The groundwaters at the Facility have a prescribed beneficial use of municipal and domestic supply in the Basin Plan. Beneficial uses must be protected for existing and probable future use.

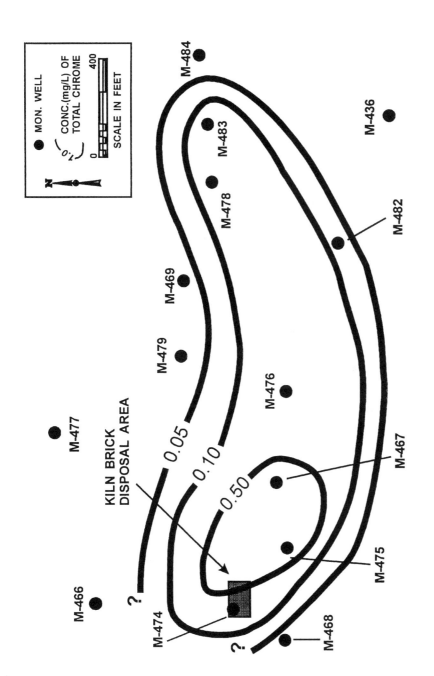

FIGURE 1

Total chrome in ground water (1990).

735

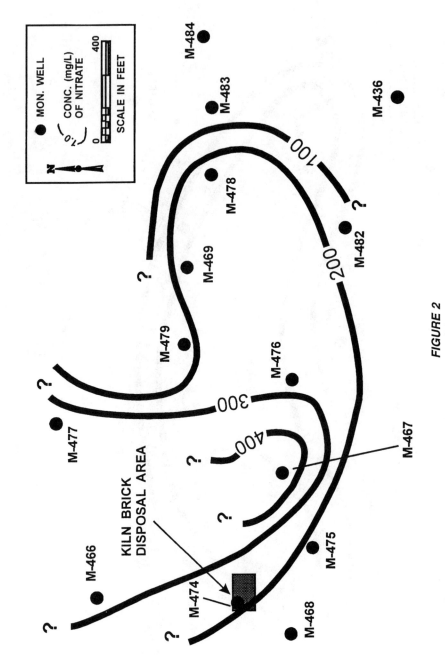

FIGURE 2

Nitrate in ground water (1990).

The presence of total chromium and nitrate at concentrations greater than the CMCLs constituted an impairment of this probable future beneficial use. This impact to a beneficial use was the driving force for the ultimate development of a cleanup strategy for the groundwater at this site.

The Basin Plan also contains a narrative goal of non-degradation. This narrative goal essentially requires responsible parties investigate any impact to back groundwater quality. The Regional Board may allow degradation of water quality when the level of degradation does not constitute an impact to a beneficial use. However, full characterization of the site was necessary before cleanup decisions could be made.

GEOLOGY

Twenty-one monitoring wells were installed in the area of the landfill between 1985 and 1994 to characterize the hydrogeology and contaminant plume. The geology in the vicinity of the landfill is variable. The sediments beneath the landfill are typically fine silts and sands. East of the landfill sediments are medium- to coarse-grained sands. The sediments in the vicinity of the landfill are underlain by weathered and fractured granitic bedrock (Figure 3). No competent low permeability strata were identified in the unconsolidated material.

As shown in Plate 1,* depth to fractured bedrock in the vicinity of the landfill is highly variable. Because the terrain is fairly flat, the resultant thickness of alluvial material generally increases to the east. The bedrock, which is high in the area of well M-476, plays an important role in hydrogeology, as discussed below.

The region upgradient of the landfill includes the quarry area for the Facility. The geology in the quarry area includes consolidated granite, limestone, and shale. Several faults transect the quarry. The faults, varying rock types, and geologic contacts influence groundwater flow and, as discussed below, also play an role in the migration of the nitrate plume.

HYDROGEOLOGY

Depth to Groundwater

Depth to groundwater is highly variable across the Facility, ranging from approximately 40 to 220 ft below ground surface (bgs) based on 1990 data. General well construction details for 22 wells in the area of the landfill are shown in Table 1.

---

*   Plate 1 appears following page 741.

TABLE 1
Monitoring Well Construction Details

Well No.	Drill Date	Depth	Screen Interval
M-436	6/82	320	120–180
M-466	1/86	100	30–90
M-467	1/86	120	40–100
M-468	8/86	95	???
M-469	8/86	167	71–131
M-474	9/86	88	45–85
M-475	9/86	98	35–85
M-476	8/86	97	44–94
M-477	9/86	170	110–165
M-478	9/86	208	120–200
M-479	9/86	151	105–145
M-482	1/87	180	85–144
M-483	2/90	232	193–232
M-484	2/90	278	226–276
M-485	2/90	162	62–162
M-487	3/91	275	215–275
M-488	3/91	111	58–111
M-489	3/91	113	57–113
M-490	11/93	110	60–110
M-491	11/93	110	60–110
M-492	11/93	270	210–270
M-493	11/93	300	240–300

Hydrographs

Between 1986 and 1992, Southern California experienced drought conditions, but during the 1992/93 winter season, the region received significant precipitation ending the drought. Water level hydrographs between 1990 and 1994 indicate different recharge behavior east and west of well M-476. Water levels, in those wells west of the bedrock high, remained relatively constant during the drought years and after the 1992/93 precipitation events. In contrast, water levels in wells east of the bedrock high declined at rates ranging from 6 to 11 ft per year during the drought years. Well M-478 went dry during this time period. Within 1 year of the 1992/93 winter season, water levels in wells to the east of the bedrock high rose to at or above pre-drought levels.

Aquifer Characteristics

The regional direction of groundwater flow generally parallels topography, which slopes to the southeast. The measured decline and rise of groundwater levels in alluvial material east of well M-476 indicate that this area is in hydraulic connection with the regional groundwater system. In contrast, the hydrograph data for wells west of well M-476 indicate that groundwater within alluvial materials of this area was not substantially influenced during times of drought and recharge. It is believed that once the groundwater level dropped below the bedrock high at M-476 (Figure 3), the groundwater west of well M-476 could only communicate with the groundwater east of M-476 through the bedrock fractures. Although the communication continued as exhibited by the constant southeasterly groundwater gradient, the fractured bedrock is not as transmissive as the alluvial materials. Thus, as represented by the lack of change in water levels, the groundwater to the west of M-476 appeared to be somewhat isolated from the seasonal influences of regional flow.

Six aquifer tests were conducted between 1987 and 1990 to evaluate aquifer characteristics. Duration of these tests ranged from 4 h to 6 d. Data from two wells screened in the alluvial material (M-467 and M-483) indicated a transmisivity ranging from 1430 to 1760 gallons per day per foot (gpd/ft). Data collected during a single test on a well screened only in the bedrock portion of the aquifer (M-485) indicated a transmisivity of 200 gpd/ft. Wells M-467 and M-485 are west of the bedrock high, and well M-483 is east of the bedrock high.

DISTRIBUTION OF CR(VI) AND NITRATE

Soil

During the initial investigation in 1986, a total of 47 soil samples were collected from monitoring well borings in the area of the landfill and submitted for laboratory analysis. Maximum sampling depth was 74 ft bgs, and 29 of the samples were collected from the upper 15 ft of alluvial material.

Chromium. Concentrations of total chromium in soil ranged from the detection limit (10 mg/kg) to 70 mg/kg. Background total chromium was determined to be below the detection limit of 10 mg/kg based on several sampling locations at the Facility away from the chromium brick disposal area. All 12 samples submitted for Cr(VI) analysis were below the laboratory detection limit (2.5 mg/kg). However, at the discretion of the sampler, all samples submitted for Cr(VI) analysis were below 15 mg/kg total chromium. The rationale for these sampling and analysis decision are unknown. It is possible that Cr(VI) was present in the soil in samples with higher total chromium levels (i.e., 70 mg/kg).

Nitrate. None of the 47 soil samples were submitted for nitrogen analysis because the source(s) appears to be upgradient and no known source of nitrate was disposed in the landfill. In order to evaluate other possible sources, samples of clay, gypsum, silica, clinker, and "background" alluvial material were collected. Nitrogen speciation was conducted by completing laboratory analyses on these samples for nitrate, nitrite, ammonia nitrogen, and total kjeldahl nitrogen (TKN).

The maximum background concentration for nitrate based on three alluvial samples was 93 mg/kg. Nitrite and ammonia nitrogen were not detected in background alluvial soils. Solubility analyses on the sampled materials indicated a maximum extractable nitrate concentration of 3.8 mg/l and a maximum extractable TKN concentration of 2.6 mg/l. Concentrations of extractable nitrate and ammonia nitrogen were near the detection limits.

Concentrations of nitrate detected in groundwater in the vicinity of the landfill exceeded 400 mg/l, as measured in 1990. Considering that the maximum extractable nitrate concentrations detected was 3.8 mg/l, it was concluded that the sampled materials were likely not the source of nitrate in the groundwater.

Groundwater

Plates 2 and 3 are charts illustrating the concentrations of total chromium and nitrate in the groundwater of wells M-489 and M-487 between 1991 and 1994. Both of these wells were used for groundwater extraction during this time period. Concentrations of nitrate generally remained constant in these two wells. However, concentrations of total chromium increased significantly in well M-489 after the winter of 1992/93.

Figure 3 illustrates the plume of total chromium in groundwater as measured in 1994. In comparison to the 1990 plume as depicted in Figure 1, the size of the plume remained relatively constant. The constant size of the plume indicates that the extraction system was successful in the objective of plume containment. However, concentrations of total chromium increased substantially near the landfill indicating that, although the extraction system was removing contaminant mass, additional contaminant mass was being solubilized into the groundwater.

DEVELOPMENT OF A REVISED CLEANUP STRATEGY

Although the groundwater extraction system appeared to be containing the contaminant plumes, the increased concentrations of total chromium and the unidentified source(s) of nitrate warranted a revised cleanup strategy. After reviewing the hydrographs and contaminant concentration data, it was concluded that total chromium concentrations generally increased consistent with precipitation events and an associated rise in groundwater levels. Water level data from monitoring

---

*   Plates 2 and 3 appear following page 741.

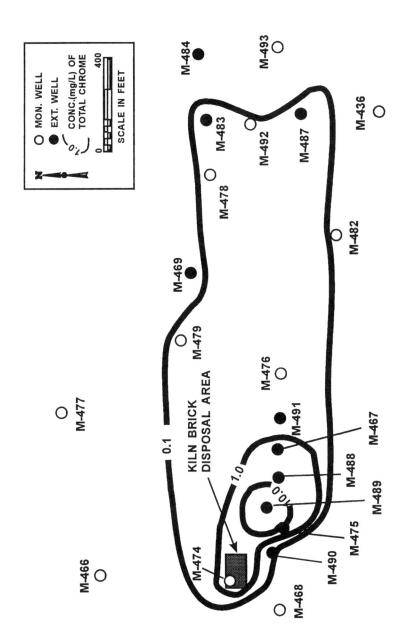

FIGURE 3

Total chrome in ground water (1994).

wells indicated that the water table was 30 ft below the base of the landfill at its highest level. Therefore, possible explanations for the total chromium increase were (1) precipitation water was infiltrating into the landfill and leaching chromium, and/or (2) groundwater was rising into chromium-contaminated vadose zone soil and remobilizing chromium. An extensive field investigation was initiated in late 1994 to find the source(s) of total chromium and nitrate in the groundwater.

Landfill Investigation

A review of available waste disposal records indicated that chromium-bearing kiln bricks were disposed in one general area of the landfill. An initial investigation of 19 borings drilled on 200 ft centers was conducted to define the limits of the brick disposal area. Borings were extended through the landfill to the underlying waste/alluvium interface. Solid samples were collected for total chromium and nitrate analysis. Solid samples in only 2 of the 19 borings contained laboratory detectable concentrations of total chromium. However, 1 of these samples contained total chromium at a concentration of 650 mg/kg. This sample was from a boring drilled immediately above the total chromium groundwater plume. Nitrate concentrations ranged from non-detected to 17 mg/kg and were typically less than 3 mg/kg. These concentrations of nitrate in soil are not indicative of a source that would generate the concentrations of nitrate that were detected in the underlying groundwater.

In order to further define the brick-disposal area, 401 soil borings were next completed on 20 ft centers. These borings were also completed through the landfill to the underlying waste/alluvium interface. In order to provide a cost savings on laboratory analysis, soil samples were first analyzed in the field by visual inspection for chromium content. Soil samples were immersed in water and allowed to decant. If the water changed to a yellow/green color, then the presence of chromium was confirmed. In-house atomic adsorption and inductively coupled plasma (ICP) analysis were conducted on some samples to confirm the presence or absence of total chromium. These in-house analyses verified that specific hues of yellow and green could be directly equated to concentrations of total chromium.

Total chromium data collected from the 401 soil borings defined the lateral extent of the former kiln brick-disposal area. A trench was then dug across the defined area to confirm the limits of kiln brick disposal, observe the overall waste content, and collect additional chromium concentration data. Waste types included kiln bricks, tires, and metal debris. Kiln bricks exhibited the yellow/green color indicative of chromium content. Other wastes adjacent to kiln bricks were stained the same yellow/green color.

Unsaturated Zone Investigation

In order to define the limits of total chromium and Cr(VI) in the unsaturated zone, a total of 12 soil borings were completed into the alluvium. Soil borings were

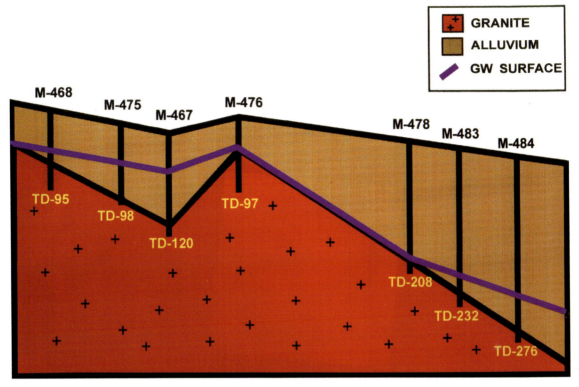

PLATE 1. East/west cross-section.

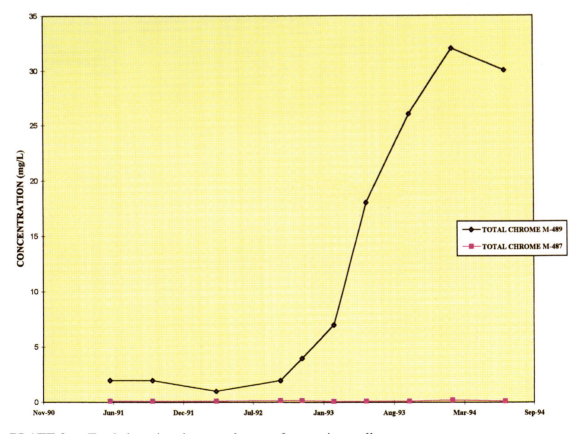

PLATE 2. Total chromium in groundwater of extraction wells.

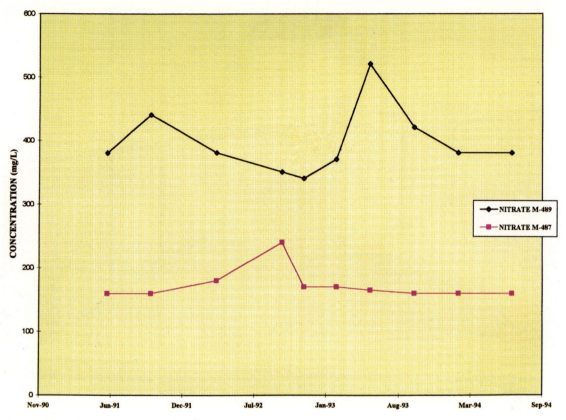

PLATE 3. Nitrate in groundwater of extraction wells.

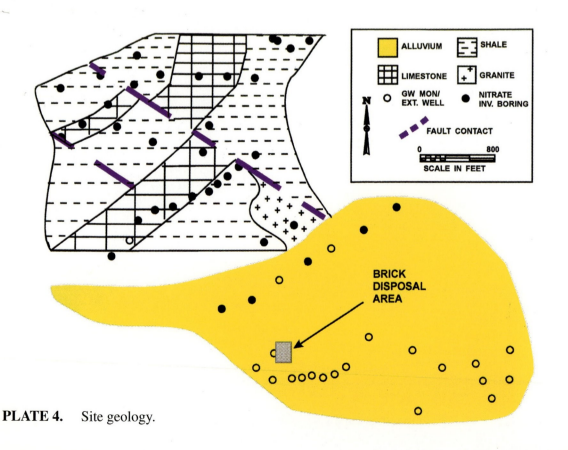

PLATE 4. Site geology.

extended vertically to bedrock, and soil samples were collected at 5-ft intervals in the unconsolidated material. All soil samples were submitted for laboratory analysis in order to accurately depict the nature of contaminants in the unsaturated zone.

Based on the laboratory analytical data, chromium migrated primarily downward through the unsaturated zone with limited lateral migration. The migration pattern is consistent with the permeable nature of the coarse grained soil beneath the landfill. The majority of chromium detected in the unsaturated zone was Cr(VI), which is also consistent with the known soluble nature of Cr(VI) when compared with Cr(III).

Nitrate Investigation

As referenced above, soil samples from the 19 borings completed in the area of the landfill during the first phase of the 1994 investigation did not indicate the presence of elevated concentrations of nitrate. Therefore, it was suspected that another source at the Facility existed. A review of records at the Facility indicated that the most likely source of nitrate would be blasting agents that consist of ammonium nitrate. It was previously suspected that some of these blasting agents may have been disposed of in the landfill. Other than the landfill, the area where these materials were used or may have been disposed was the quarry, which is hydrologically upgradient of the landfill. The other possible source identified was naturally existing nitrogen, which could be present in the metamorphic rocks of the quarry.

Quarry Investigation. In the early stages of ammonium-nitrate/fuel oil (ANFO) blasting technology (1950s to early 1960s) blast holes that encountered groundwater were loaded with manufactured charges of ANFO. The loaded charges were sealed in plastic bags with a fibrous cover to prevent saturation prior to detonation. Because these charges were less dense than water, wooden poles were used to submerge them. Each hole required multiple charges, and some charges were occasionally punctured by the wooden poles during the loading process. ANFO will not detonate once saturated, and thus ammonium nitrate was in some cases left below the quarry floor within the groundwater.

In early 1995 an investigation was undertaken in the quarry area upgradient of the landfill. The investigation consisted of drilling 35 holes through competent bedrock using a duel-wall reverse circulation drilling rig. Water samples were collected from each boring after purging for 5 to 10 min. Field data collected from each boring included the depth to the first encountered water level, static water level, and rock type. Water samples were filtered and submitted for nitrate analysis. Plate 4* illustrates the geology of the quarry area as determined from new and existing data. Figure 4 illustrates the groundwater flow patterns and nitrate concentrations.

---

* Plate 4 appears following page 741.

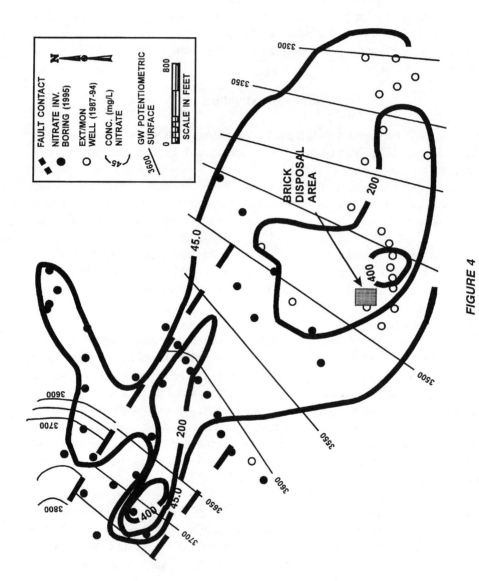

FIGURE 4

Nitrate in ground water (1994).

Groundwater samples collected from quarry borings appear to indicate three specific plumes of nitrate groundwater degradation, with a maximum detected concentration of 528 mg/l. These three plumes were traced to known areas of undetonated blasting materials.

By comparing Plate 4 and Figure 4, it is noted that the contaminant migration patterns are controlled by the local geology. The high hydraulic conductivity of the limestone that is present in much of the source area promotes movement of the plumes into the adjacent alluvium. The southeast trending faults restrict movement of the plume to the east and southwest. The low hydraulic conductivity of the shale unit at the eastern edge of the source area restricts movement of the plume in this area.

In summary, nitrate appears to have leached from ANFO. By collecting grab groundwater samples from borings (Figure 4), the sources of nitrate were traced to three specific areas where blasting problems previously occurred and the ANFO was reportedly left in place (Figure 4). The migration of the plume from the quarry area has been controlled by local geology, including geologic contacts and fault structures. Although ANFO has not been found in the landfill, the nitrate groundwater plume geometry (Figure 4) and the elevated concentrations of nitrate in the groundwater in the immediate vicinity of the landfill suggest that ANFO was disposed of in the landfill.

Naturally Existing Nitrogen. Samples of cuttings from the shale hardrock borings were collected and submitted for laboratory analysis for TKN, nitrite, ammonium nitrogen, and nitrate. These data indicate extractable TKN concentrations ranging from 100 to 190 mg/l. Nitrate ranged from non-detected to 21 mg/kg. The presence of extractable TKN indicates the presence of a significant amount of natural organic nitrogen within the metamorphic rock. Under oxidizing conditions the TKN could be transformed to nitrate.

The contribution of natural organic nitrogen to the entire plume of nitrate is difficult to quantify. It is known that some of the groundwater that is found in the alluvial material near the landfill must first pass through the upgradient shale units. Based on water quality data collected in the shale unit during the early 1995 investigation, background nitrate in the shale unit is estimated to be 60 mg/l. Although cleanup levels for nitrate have yet to be determined, the existence of natural organic nitrogen will likely play an important role in this process.

REVISED CLEANUP STRATEGY

Based on the findings of the 1994 and 1995 field investigations, it was evident that sources of chromium and nitrate remained at the Facility. Unless the sources were removed or controlled, the groundwater extraction system would only contain the plume. Other remedial actions were clearly necessary to accomplish source removal, which were not a primary focus of the groundwater extraction system. Furthermore, based on the rise in total chromium concentrations near the landfill in 1993 and 1994, the concentrations could be expected to rise again during further precipitation events.

Removal of Chromium Source

The chromium source removal involves the excavation of chromium bearing bricks and unsaturated zone soil contaminated by chromium. The bricks and soil will be used as source material for the normal industrial processes at the Facility. The excavated area will be back-filled with shale and covered with a monolithic cover of 1×10^{-5} cm/s native soil to limit the infiltration of precipitation water. The cover will be graded and maintained to prevent ponding of rainwater. The monolithic nature of the cover system will promote evaporation of precipitation. Vegetative growth that naturally develops on the native soil cover further limits infiltration of precipitation by transpiration processes. The entire excavation, recycling, back-fill, and cover installation process was initiated in late 1995 and is expected to extend into 1997.

The cleanup level for Cr(VI) in unsaturated zone soil was developed based on backgroundwater quality and a generic site-safety factor. The goal is not to fully eliminate the contact of precipitation with waste, but rather to limit the potential for any leachate to migrate to the underlying groundwater.

The site-safety factor approach is often referred to in California as the *designated level* methodology, and it is intended to depict the potential for leachate generation and transport to the underlying groundwater. Rather than conduct site-specific leaching analysis, the safety factor is typically developed in consideration of rainfall, geology, depth to groundwater, and known contaminant characteristics. This method can be conservative because if a site-safety factor of one is used it will then be assumed that all available contaminant mass is leachable and will be transmitted to the groundwater. However, this method also invokes reason in that a higher safety factor can be used to reflect that all contaminant mass is rarely leachable and, depending on site geology, all leachate generated will not be transmitted to the underlying groundwater.

The background concentration of total chromium in groundwater is 0.03 mg/l based on a well upgradient of the landfill. In consideration of the permeability of the native soil cover, rainfall of less than 8 in per year, the positive drainage of the cover materials, and the depth to groundwater, a site safety factor of 100 was agreed on. The resultant cleanup level of 3.0 mg/kg for Cr(VI) was then developed by multiplying the backgroundwater quality concentration by the site-safety factor.

The cleanup level assures several things. First, contaminated soil within 20 ft of the ground surface will generally be removed. Therefore, any precipitation that infiltrates through the cover system will need to percolate 20 ft before contacting any contaminated soil. Second, any precipitation water that contacts contaminated soil will have a very low mass of contaminants from which to generate leachate.

Any leachate that may develop would be low in concentration of chromium VI and would still need to percolate at least an additional 20 ft before contacting groundwater.

Disposal of Kiln Brick and Contaminated Soil

Excavated kiln brick and contaminated soil will be used as process material at the Facility. The volume of brick and soil is minimal in comparison to the volume of limestone ore processed at the Facility. The chromium becomes a part of the final product, which is portland cement. Sampling of the final product on a regular basis has confirmed that chromium concentrations in the portland cement are typically at or near detection limits for chromium. Aside from the useful introduction of alumina, the final product is essentially unaffected by the introduction of brick into the process.

Removal of Nitrate Source

The majority of the nitrate source is suspected to be in the quarry area. The affected areas of the quarry are planned to be mined during the next several years. As these areas are mined the undetonated blasting materials are removed. This mining is expected to extend until the year 1999.

There are no existing groundwater extraction wells in the quarry area. The extraction system in the area of the landfill downgradient of the quarry is currently capturing nitrate-effected groundwater emanating from the quarry. As blasting materials are removed the existing extraction system will be evaluated for efficiency in lowering nitrate concentrations. Extraction wells in the nitrate source areas may be necessary at a later date based on further evaluation of the extraction system.

Based on the shape of the nitrate plume, one could argue that the landfill still appears to be one source of nitrate to the groundwater. No nitrate was detected in landfill materials or unsaturated zone soil beneath the landfill. Furthermore, no blasting materials or other possible sources of nitrate were noted during trenching of the landfill. Despite these findings, additional landfill investigation may be necessary at a later date should groundwater pumping not be effective in reducing nitrate concentrations in the groundwater in the vicinity of the landfill.

FINANCIAL CONSIDERATIONS

Data collected at this Facility further solidifies the author's belief that groundwater extraction systems are generally useless in reaching cleanup goals unless the source(s) of the contaminant mass are removed. The estimated annual cost for operation and maintenance of the 11 well extraction system is $20,000. Whether the goal of the extraction system is solely plume containment or containment and restoration of groundwater quality, the costs and associated liability could theoretically continue forever without mass source removal. Although reaching cleanup goals will still be a timely process, removal of mass sources may result in the goals being feasible. With this concept in mind, a decision was made by the Facility to locate and remove contaminant sources.

In order to be cost-effective, staff at the Facility completed much of the above detailed 1994 and 1995 investigations using internal resources. On-going and future remediation activities will also be completed using mostly internal resources. The following is a brief discussion of the cost-effective nature of the investigation and cleanup.

Soil Borings

Drilling of the soil borings in the landfill area were completed using quarry exploration equipment owned by the Facility. A California registered geologist (CRG) is on staff at the Facility and was present to log each hole. The 401 hole landfill drilling program took only 10 working days.

The 36 borings completed in the quarry area during the nitrate investigation were drilled using a outside contractor. The CRG logged each of the holes and collected all of the samples. The drilling program lasted 9 d. The cost of contracting the drilling rig and crew is estimated at just under $26,000.

Laboratory Analysis

As discussed above, field evaluation of soil samples was used to limit the number of samples submitted for laboratory analysis. In-house laboratory equipment was also used to limit the number of samples submitted to an outside laboratory.

Water Treatment

As of 1994 approximately 7.5 million gallons of groundwater had been extracted and treated. As discussed above, water is used in the industrial process at the Facility. The cost of treatment consisted only of the extraction system design, construction, and operation.

Kiln Brick and Contaminated Soil

The excavation of kiln brick and contaminated soil is being completed by staff of the Facility. Two staff received 40-h hazardous waste materials training specifically for this task. Estimated staff time for the project is 200 man days. Two part-time staff were hired to conduct regular activities at the Facility during the excavation project.

In order for the material to be processed at the Facility, a significant amount of sorting must be completed first. The sorting is necessary because the Facility's primary crusher for limestone ore cannot receive metal debris or other wastes that could potentially cause damage. Most of the sorting equipment was purchased specifically for this project at a cost of approximately $300,000. This equipment includes vibrating grizzlies, magnetic belts, and a jaw crusher. The cost of this equipment is significantly less than the cost for disposal of the estimated 50,000 cubic yards of material at an appropriate off-site waste disposal/treatment facility. The estimated cost for disposal of 50,000 cubic yards of chromium contaminated soil and kiln brick is in excess of a million dollars.

Total Costs

The estimated total costs for the investigation and cleanup is $750,000. If the investigation and cleanup were completed using outside contractors and consultants, the costs would obviously have been substantially higher. Based on experience with other sites with similar soil and groundwater problems, the estimated cost of outside contractor services would have been in excess of 4 million dollars. Thus, the costs of a cleanup the magnitude of this project could have been prohibitive without the use of internal resources.

CONCLUSIONS

A groundwater extraction system was installed to remove contaminant mass and control plume movement. However, failure to remove contaminant sources resulted in increased contaminant concentrations.

By reviewing hydrographs, contaminant concentrations over time, and available information regarding past disposal practices, conclusions were drawn regarding necessary actions to restore water quality. Intensive field investigation provided the necessary data to design a revised corrective action program that combines source removal and continued groundwater extraction and treatment.

By using internal resources, staff at the Facility were able to control costs for the project. The actions of staff at the Facility should be a model for other sites where restoration of water quality is technically feasible but economically infeasible.

REFERENCES

All technical reports utilized for this paper are unpublished public records. These records are available for review at the office of the California Regional Water Quality Control Board, Victorville Office.

Journal of Soil Contamination, 6(6):751–765 (1997)

Effects of Chromite Ore-Processing Residue on Concrete Structures

A. Rhett Whitlock and Christopher L. Galitz

Whitlock Dalrymple Poston & Associates, Inc., 8832 Rixlew Lane, Manassas, Virginia 20109

In the 1960s, a number of industrial and commercial buildings in the Kearny, New Jersey, area were constructed on or adjacent to locally produced chromite ore-processing residue (COPR). This residue, which has the properties of a sandy/silty soil, was used as a fill material to reclaim the indigenous swampy ground conditions, known locally as the "Meadowlands". Generally, the fill stratum is between 4 and 10 ft thick and contains varying amounts of hexavalent and trivalent chromium. It has been alleged that chromium causes damage to structures through various mechanisms. The authors have considered these allegations through research and review of the literature, field investigations of over 100 structures built on or near COPR, field studies involving concrete exposed to COPR, and laboratory studies. The results of the investigations and studies are discussed in this article. It has been found that the existing distress on building and bridge structures is no different from the types and magnitudes of distress seen on other similar structures founded on non-COPR soils. The distress observed was attributable to such things as normal weathering, impact, lack of maintenance, improper design or construction, and application of deicer salts. The field experiments for the effects of COPR on concrete have shown no correlation between the level of hexavalent or trivalent chromium and the deterioration of concrete. The single most important factor to the deterioration of concrete was the quality of the concrete mixture. Similarly, the laboratory experiments on corrosion of reinforcing steel in concrete or soil showed no effects of COPR on corrosion.

1058-8337/97/$.50

INTRODUCTION

*I*N Hudson County, New Jersey, various industrial concerns processed chromite ore from the turn of the century until approximately 1971. A byproduct of the process was chromite ore-processing residue (COPR) that looks like and has similar construction use characteristics to a sandy/silty soil. Because of these characteristics, this material (that contained residual chromium in the trivalent and hexavalent forms) was used as "fill" on industrial, commercial, and residential sites (Sheehan, *et al.*, 1991). The chromite chemical product plants in Hudson County, New Jersey generated approximately 2.75 million tons of the COPR materials (MEMT, 1990).

The Hudson County area contains a large number of industrial and commercial business locations, most of which were constructed in the 1950s and 1960s. At that time, the available land was a low-lying marsh area, known as the Meadowlands; and, therefore, in order for construction to take place, the sites needed to be filled. The local availability of COPR, with its soil-like properties, made its use as a fill material to reclaim the low-lying areas logical and economical. The depth of fill varied from site to site, but was generally less than approximately 10 ft (3 m). Over the years, some structures built on or adjacent to COPR experienced problems which were alleged to be due to the fill material.

Problems alleged to be associated with the use of COPR as a fill material include strength degradation of concrete (Huang, *et al.*, 1994), loosening of mortar bonds to brick (Raghu and Hsieh, 1989a), and displacements of concrete floors, walls and columns due to expansion of the residue (Raghu and Hsieh, 1989b). Recently, a comprehensive effort has been made to examine these allegations and to determine, based on literature review, laboratory and field testing, and study of existing concrete structures, the effects, if any, of COPR on concrete and concrete structures. This article presents the results of these efforts to date.

LITERATURE REVIEW

Little research has been published regarding the effects of COPR, and more specifically, hexavalent chromium, on the strength characteristics of concrete. It has been reported that the strength of concrete is decreased by the addition of potassium chromate, a corrosion inhibitor, to the concrete mix (Craig, 1969). This decrease, relative to control batches in which no additive was used, was largest at early stages (<28 d), and became less significant as the age of the concrete increased. Craig's data showed that the largest decrease in strength at 28 d was in the batch containing the least amount of additive (2%). At 150 d, the strengths of the three batches with additive (2, 4, and 6%) were very nearly equal. The 28-d strength of the 2% additive batch was approximately 67% of the strength of the control batch. At 150 d, the strength of the 2% batch was approximately 76% of the control batch. Craig also reports a strength reduction with other chemicals used as corrosion-inhibiting admixtures. The decreases in 28-d compressive strength for potassium chromate, sodium nitrite, and sodium benzoate were approximately 33, 20, and 40%, respectively (Craig, 1969).

It has also been reported that the adsorption of hexavalent chromium by concrete particles causes a degradation of concrete strength (Huang *et al.*, 1994). This is stated as a possibility when concrete is exposed to acidic chromium solutions for long periods of time. According to Eglinton (1987), this would fall under the category of acid attack, and occurs when concrete is exposed to an acidic solution, (i.e., pH less than approximately 6). Data collected to date indicate that, in general, COPR soils in the area had pH values greater than 8, with an average of approximately 11.5. Non-COPR soils showed pH values of approximately 6.5. According to Weng *et al.* (1996), the adsorption that may lead to loss of concrete's buffering capacity, allowing acid attack, did not occur except within the pH range of approximately 3 to 8. Concrete samples from bridge piers in the northern New Jersey area, exhibited pH measurements between approximately 7.5 and 8.9, where the effects of adsorption approached minimal levels (Poston *et al.*, 1995). Based on pH values observed both in COPR soils, and those of in-service concrete, it is not likely that the adsorption of hexavalent chromium would lead to acid-related degradation of the strength of concrete or the integrity of concrete structures.

Another mechanism alleged to be responsible for damage to structures constructed on chromite ore-processing residue is the crystallization of chromium salts within a material. It is the opinion of one set of authors that this is the mechanism responsible for alleged expansion of COPR fill (Raghu and Hsieh, 1989b). Further, they report this crystallization process to be responsible for the strength degradation observed by Craig. It is their opinion that cementitious materials (concrete and mortar) in contact with chromium-bearing soils lose strength due to the permeation of chromium salt solutions into the material and subsequent evaporation of the water. This process leaves behind chromium salt crystals that expand as they crystallize. In the case of mortar, Raghu and Hsieh allege that the chromium salts cause the mortar to soften and flake, eventually losing the ability to adhere to brick units in a wall, causing "brick fallout" (Raghu and Hsieh, 1989b).

In a concrete bridge pier, chromium salts were reported to have migrated up to 8 ft (8') vertically from grade level (Raghu and Hsieh, 1989b). Poston, et al conducted a survey of 15 bridges in New Jersey built on chromium sites (Poston *et al.*, 1995). Their investigation revealed no evidence of chromium salt crystallization in any of the over 30 concrete bridge piers examined.

There is some discrepancy within the literature as to the mechanisms, if any, responsible for strength degradation of concrete exposed to COPR. None of the observations or conclusions reviewed were based on laboratory testing or other scientific testing of concrete after exposure to a field environment containing COPR. Field observations have been cited by other researchers; however, the buildings examined numbered less than 10.

The authors have inspected over 100 buildings in the Kearny, New Jersey, area to determine the cause(s) of deterioration, if any, and to investigate the possible link between the presence of COPR as a fill material and deterioration of building structures. In addition, over 30 bridges in the area were inspected with a similar purpose. The buildings investigated varied in type of construction, but when taken in total encompass nearly the full range of common building materials in use today, that is, brick masonry, concrete masonry, steel, timber and concrete. The bridges examined were constructed primarily of reinforced concrete.

The most common building material seen in exterior walls was concrete masonry. In general, damage to these walls included shrinkage cracking, settlement-induced cracking, impact damage, and deterioration caused by freeze/thaw action. Plate 1 shows a typical concrete masonry wall and the deterioration of its face due mainly to freeze/thaw action. This condition was seen in a large number of locations, generally below leaky gutters and downspout, or at exterior grade. The severity of the damage was directly a function of the continuous exposure to water and in no case related to the presence of COPR fill materials. In most cases, this form of deterioration was most severe toward the tops of the walls, far above the level of potential migration of chromium salts. Cracking and impact damage seen in exterior concrete masonry unit (CMU) walls was common throughout the area, independent of the presence or absence of COPR fill.

In general, brick masonry walls examined as part of our investigations showed similar cracking to that seen in CMU walls. This deterioration was mainly caused by settlement. In addition, there were locations where deterioration, i.e., corrosion, of the steel supports and reinforcement had caused damage to the brick facade. Plate 2 shows cracking and displacement of a brick masonry wall just above a doorway. There was no evidence of chromium salts at this location, suggesting that the damage observed was fully independent of the interaction of COPR with the building materials. It was concluded based on observations and experience with similar occurrences on other non-COPR projects that the damage seen was caused by severe corrosion of the door lintel, accompanied by expansion of the steel, and, therefore, displacement of the brick.

In general, the concrete seen at the various sites was used for foundation construction, slabs-on-grade, and retaining walls. At some sites, above-grade concrete foundation walls were used to support raised loading dock floors. Finally, at the bridge sites concrete was used for foundations and pier bents supporting raised roadways and bridges. In all cases where there was evidence of significant deterioration of concrete, it was attributed to corrosion and subsequent expansion of reinforcing steel and/or freeze/thaw deterioration. Plate 3 depicts severe deterioration of a bridge support due to corrosion. The site that houses this bridge contains

---

*   Plates 1 through 4 appear following page 759.

COPR as a fill material. However, the deterioration shown in the photograph is approximately 15 ft above grade and therefore cannot be due to the presence of chromium. The area of severe damage to the bridge support was directly in line with a drainage spout from the roadway above. Chloride ion testing in this area showed elevated levels, suggesting that the catalyst for the corrosion was not the presence of chromium in the soils of the site, but chloride de-icers used on the roadway which drained down onto the reinforced concrete (Poston *et al.*, 1995).

During our site investigations other damage, such as slab displacement and wall rotation, was noted. The main reason for the damage was excessive movement of the walls and superstructures of some buildings caused by differential settlement. Plate 4 shows an area where the concrete floor slab of a warehouse building shows evidence of "humps" at various locations along its edge.

Examination of site geology reveals that the fill soils are underlain by a layer of highly organic soils resembling peat moss (Whitlock and Moosa, 1996). This soil stratum is the remnants of the once surface marshy soils and is typically referred to as the meadow mat layer. Due to its highly organic nature, it undergoes long-term consolidation, allowing settlement of the fill layer above, and therefore any structure built on this fill layer. Figure 1 shows how the consolidation of the meadow mat layer allows the settlement of a floor slab, while the pile foundation, founded in soils below the meadow mat, remains relatively stationary. The resulting differential settlement causes cracking and displacement of the slab. There were many examples of this settlement pattern seen throughout buildings in the Hudson County, New Jersey area.

Some causes of deterioration were noted at several sites, regardless of construction type or materials. These causes include improper design, faulty construction, and a general lack of maintenance of the buildings. Design problems included poor

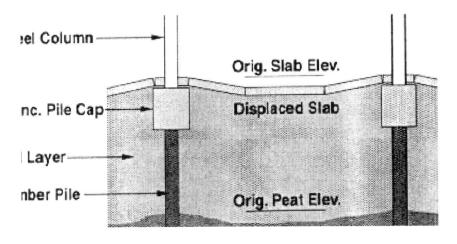

FIGURE 1

Differential settlement caused by consolidation of the meadow mat layer.

detailing of parapet wall connections to the superstructure, improper foundation reinforcement, and the lack of proper control and expansion joints in masonry walls. Each of the factors has contributed to damages observed in the Hudson County area, to one degree or another. Construction deficiencies include improper placement of reinforcing steel within foundation walls and improper fill procedures (no compaction testing and organic materials used as fill). Finally, it was seen on nearly every building that the lack of maintenance of control and expansion joints, gutters and down spouts, wall coatings, and caulking joints have allowed excessive water penetration of exterior walls and openings. This water penetration has caused severe corrosion and freeze/thaw damage at several locations.

LABORATORY TESTING

Corrosion

During our site investigations, it was observed that the deterioration of concrete structures was due in part to corrosion of reinforcing steel. To examine the effect, if any, of COPR and/or hexavalent chromium in the form of chromates on the corrosion process, the corrosion potential of COPR soils and chromate solutions on steel bars and reinforced concrete, respectively, was tested (Poston et al., 1996). Figure 2 shows a typical concrete "cell" on which chromate solutions, in varying concentrations, were ponded. The current induced by corrosion of the steel embedded in the concrete was measured periodically. These readings were compared with similar readings for cells subjected to water and those subjected to salt. The results, shown in Figure 3, indicated that hexavalent chromium, in the form of chromates, did not affect the corrosion potential of steel reinforcing within concrete. Similar cells, constructed using COPR soils instead of concrete, were tested

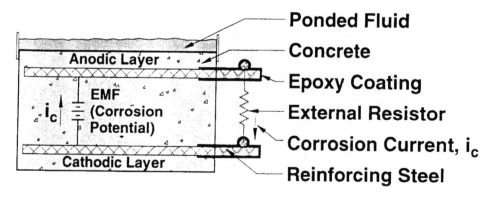

FIGURE 2

Typical reinforced concrete cell used for corrosivity testing.

756

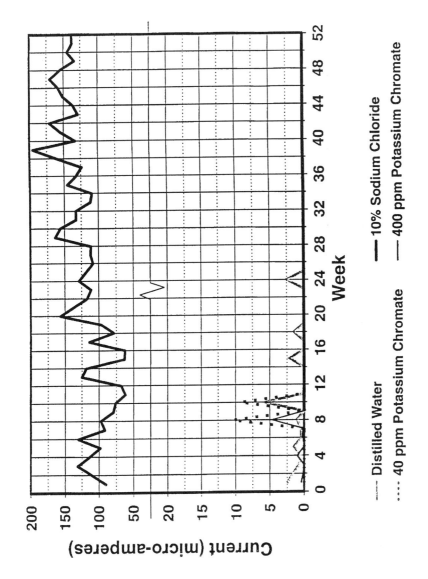

FIGURE 3

Average corrosion current in laboratory concrete specimens.

for the same 1-year duration. As with the concrete cells, results were compared with those of control cells and demonstrated that COPR does not affect the corrosion potential of steel.

Structural Integrity

As part of site investigations and consideration of the condition of concrete within the structures examined, the authors collected concrete core samples from the floor slabs, foundation walls, and footings of several sites. The concrete samples varied in age between approximately 19 and 50 years. The majority of the concrete cores were tested for compressive and splitting tensile strength with selected samples being submitted for petrographic and chemical analysis. In general, results showed the concrete to be structurally sound with compressive strengths significantly higher than original design strength. The microscopic examination performed as part of petrographic analysis revealed no significant mechanisms for loss of structural integrity of the concrete. In general, deposits of ettringite, a byproduct of chemical reactions in the concrete and long-term exposure to moisture, were seen, as expected. These reactions and their byproducts were not linked to the presence of COPR. Finally, chemical analysis revealed that hexavalent chromium had in fact infiltrated the concrete. Therefore, it was seen that the concrete had been subjected to a COPR environment, but, there were no identified effects on the concrete due to COPR.

A long-term field and laboratory testing protocol to study the effects of high chromium environments on concrete cylinders when compared with a low chromium field environment and laboratory control specimens was developed. This study involved the manufacture of over 600 concrete samples of varying properties; placement of these cylinders into two (2) burial sites in the Hudson County, New Jersey, area; removal of the samples at varying time intervals; physical, chemical, and petrographic testing of those cylinders; and statistical and engineering analyses. The study was implemented in November of 1992 when a total of four hundred forty (440) cylinders were placed in COPR fill areas at two different sites, one with a high chromium content in the soils (>100 ppm Cr(VI)) and one with a low chromium content (<10 ppm Cr(VI)). The cylinder specimens were of four types of concrete, varying in quality, as measured by cement content and air entrainment, and type of aggregate. At various time intervals, ranging from 30 d to 3 years, cylinders were extracted from the field and tested. Results were compared with those of identical laboratory control specimens that were maintained at standard lab environment conditions. Results of the three exposures were compared statistically and trends in the data over time were considered.

The results of the compressive strength testing are shown in Figure 4, where strengths have been normalized by dividing by the average strength of laboratory control samples of the same batch. The three exposures were compared statisti-

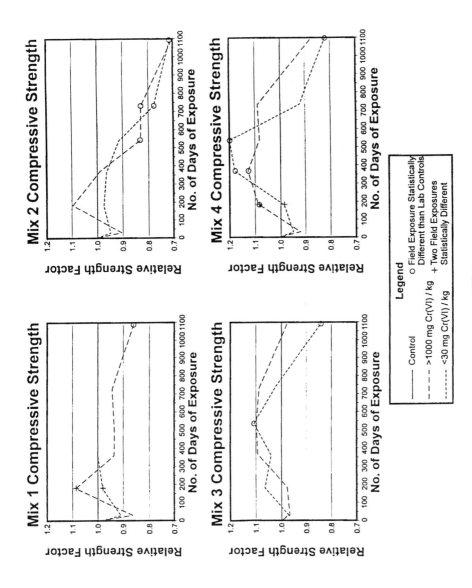

FIGURE 4

Compressive strength of concrete exposed to COPR environments.

759

cally, at a 5% level of significance, with significant differences shown on the graphs of Figure 4. It can be seen from these graphs, and the points denoting statistical differences, that the high and low chromium environments have followed similar trends over time. Although field exposure specimens tended to have lower strengths than those of laboratory controls, the differences between the strengths of specimens from the two different exposures are not significant, and therefore the apparent loss of strength cannot be linked to the concentrations of chromium in the exposure soils.

The results from the authors' testing program, as well as those from field-collected concrete cores, were considered using the ratio of actual strength to design strength as a comparison factor. Pertinent compressive strength results, with the design strengths, for each sample are given in Table 1. Note that for the oldest sample, the design strength was unknown; therefore, based on knowledge of similar construction types for the Hudson County, New Jersey, area, a conservative value of 3500 psi was assumed. The results of the comparison are shown graphically in Figure 5. In all cases, the actual strengths of the concrete were higher than design strengths. Further, there is no trend in the data that would suggest that concrete loses strength over time due to exposure to COPR soils.

On-Going Work

The analyses conducted to date by the authors have focused on the effects of COPR on cast-in-place concrete. Those effects were examined using field samples exposed to high chromium environments for long periods of time and laboratory-fabricated samples exposed for short periods of time. The testing program in place is being extended to produce results over the course of the next few years to assess any effects that COPR may have on concrete.

CONCLUSIONS

Over 100 buildings and 30 bridge structures were examined in the northern New Jersey area. In no case has observed damage been directly attributed to the possible use of COPR as a fill material. Damages that were observed in buildings were related to common problems, such as concrete shrinkage, freeze/thaw deterioration, differential settlement, improper design/construction, or lack of appropriate maintenance. In bridges, deterioration was due, in most part, to corrosion of reinforcing steel and/or alkali-silica reactivity.

Laboratory studies have been conducted that show that chromium and COPR soils do not increase the corrosion potential of embedded steel in concrete and soil samples, respectively. Multiple concentrations of hexavalent chromium were considered.

PLATE 1. Deterioration of a concrete masonry wall.

PLATE 2. Cracking and displacement of a brick masonry wall.

PLATE 3.　Deterioration of a bridge pier due to corrosion of reinforcing steel.

PLATE 4.　Differential settlement of a warehouse floor slab at pile caps.

TABLE 1
Comparison of Strength Testing to Design
Strength of Concrete Exposed to COPR Environments

Sample type	Age at test (years)	Design strength (psi)	Actual strength (psi)		Actual/design strength ratio	
			Exposed to COPR	Control	Exposed to COPR	Control
Lab 1	0.08	2500	3250	3770	1.30	1.51
Lab 3	0.08	2500	4160	4600	1.66	1.84
Lab 6	0.08	4500	7740	8010	1.72	1.78
Lab 8	0.08	4500	6870	7450	1.53	1.66
Lab 1	0.50	2500	3590	3290	1.44	1.32
Lab 3	0.50	2500	4120	3760	1.65	1.50
Lab 6	0.50	4500	6850	6990	1.52	1.55
Lab 8	0.50	4500	7680	7050	1.71	1.57
Lab 1	1.00	2500	3270	3490	1.31	1.40
Lab 3	1.00	2500	3940	3970	1.58	1.59
Lab 6	1.00	4500	8060	7380	1.79	1.64
Lab 8	1.00	4500	7440	6590	1.65	1.46
Lab 1	1.50	2500	2950	3150	1.18	1.26
Lab 3	1.50	2500	3250	3910	1.30	1.56
Lab 6	1.50	4500	7590	6900	1.69	1.53
Lab 8	1.50	4500	6900	6370	1.53	1.42
Lab 1	2.00	2500	2910	3090	1.16	1.24
Lab 3	2.00	2500	3380	4070	1.35	1.63
Lab 6	2.00	4500	7680	7050	1.71	1.57
Lab 8	2.00	4500	7480	6870	1.66	1.53
Lab 1	3.00	2500	2910	3380	1.16	1.35
Lab 3	3.00	2500	3080	4350	1.23	1.74
Lab 6	3.00	4500	6770	6920	1.50	1.54

TABLE 1 (continued)
Comparison of Strength Testing to Design
Strength of Concrete Exposed to COPR Environments

Sample type	Age at test (years)	Design strength (psi)	Actual strength (psi)		Actual/design strength ratio	
			Exposed to COPR	Control	Exposed to COPR	Control
Lab 8	3.00	4500	5900	6750	1.31	1.50
Floor slab	17.00	3000	5750		1.92	
Pile cap	28.00	2500	5590		2.24	
Pile cap	28.00	2500	5430		2.17	
Pile cap	28.00	2500	5920		2.37	
Foundation wall	28.00	2500	5300		2.12	
Foundation wall	28.00	2500	5950		2.38	
Foundation wall	28.00	2500	4190		1.68	
Floor slab	28.00	3000	8000		2.67	
Floor slab	28.00	3000	9400		3.13	
Floor slab	29.00	2500	3810		1.52	
Floor slab	29.00	2500	2920		1.17	
Floor slab	29.00	2500	4400		1.76	
Footing	29.00	2500	3790		1.52	
Footing	30.00	2500	3750		1.50	
Footing	30.00	2500	4890		1.96	
Floor Slab	30.00	3000	5450		1.82	
Floor Slab	50.00	3500[a]	4290		1.23	

[a] Assumed value.

762

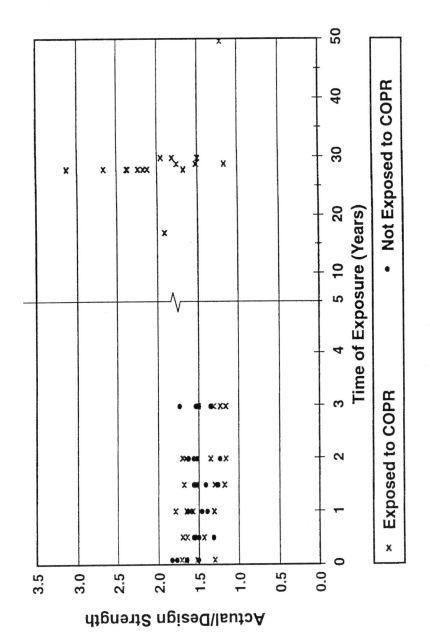

FIGURE 5

Comparison of strength testing to design strength of concrete exposed to COPR environments.

A large-scale field and laboratory testing program has been implemented to track the strength performance of concrete samples over time, while subjected to high and low chromium COPR environments. The results to date, encompassing 3 years of testing, show that there is no significant difference between the two exposure types, and therefore, that any strength degradation observed in the field is not due to chromium.

Testing of concrete samples that had been exposed to COPR soils for 20 to 50 years showed that in no case did the exposure to chromium cause degradation of concrete strength or the performance of concrete structures.

Over the course of the authors' investigations, samples and structures exposed to COPR were compared with those not exposed to COPR. In all cases, deterioration was seen to be similar in both sets of observations. From this, it was seen that concrete does not experience any deleterious effect due to exposure to chromium in COPR.

Finally, the literature shows that many organizations and individuals have classified COPR, and, specifically, hexavalent chromium as being hazardous to the integrity of concrete structures. This opinion can not be supported or endorsed by the authors based on studies completed to date. We have conducted multiple studies, investigated over 100 structures, and considered all literature on the subject and are of the opinion that the possible presence of COPR per se as a fill material does not have any significant effects on the integrity of concrete or concrete structures.

REFERENCES

Chromium Producing Industry, Public Health Service Publication No. 192, Public Health Service, 1953.

Craig, R. J., Physical Properties of Cement Paste with Chemical Corrosion Reducing Admixtures, M.S. Thesis, Submitted to Purdue University in Partial Fulfillment of Requirements for the Degree of Master of Science in Civil Engineering, June 1969.

Eglinton, M. S. 1987. *Concrete and Its Chemical Behavior,* London, England, Thomas Telford Ltd.

Huang, C. P., Weng, C. H., Allen, H. E., and Cheng, A. H.-D. Effect of Specific Chemical Reaction on the Transformation and the Transport of Chromium in the Soil-Water System and in Concrete Walls, Report submitted to the New Jersey Department of Environmental Protection and Energy, August 1994.

Montclair Environmental Management Team (MEMT). 1990. International Journal of Environmental Studies, Vol. 35, pp. 263–275.

Poston, R. W., Whitlock, A. R., Galitz, C. L., and Kesner, K. E. 1995. Evaluation of Bridges in Chromite Ore Processing Residue, National Research Council, Transportation Research Board, Record 1460, June 1995.

Poston, R. W., Galitz, C. L., and Yates, J. S. 1996. Corrosion susceptibility from chromium exposure of steel in concrete and soil. Pending publication in *ASTM Journal of Testing and Evaluation.*

Raghu, Dorairaja and Hsieh, H.-N. 1989a. Performance of some structures constructed on chromium ore fills, *J. Perform. Construct. Facil.* **3(2),** 113–120.

Raghu, Dorairaja and Hsieh, H.-N. 1989b. Origin, properties and disposal problems of chromium ore residue, *Int. J. Environ. Studies* **34,** 227–235.

Sheehan, P. J., Meyer, D. M., Sauer, M. M., and Paustenback, D. J. 1991. Assessment of the Human health risks posed by exposure to chromium-contaminated soils, *J. Toxicol. Environ. Health* **32,** 161–201.

Weng, C. H., Huang, C. P., Allen, H. E., Leavens, P. B., and Sanders, P. F. 1996. Chemical interactions between Cr(VI) and hydrous concrete particles. *Environ. Sci. Technol.* **30,** 371–376.

Whitlock, A. R. and Moosa, S. S. 1996. Foundation design considerations for construction on marshlands. *J. Perform. Construct. Facil.,* **10(1),** 15–22.

Journal of Soil Contamination, 6(6):767–797 (1997)

Traditional and Innovative Treatment Methods for Cr(VI) in Soil

Thomas E. Higgins,[1] Amy R. Halloran,[2] and John C. Petura[3]

[1]CH2M HILL, Reston, VA; [2]CH2M HILL, Albuquerque, NM; [3]Applied Environmental Management, Inc., Malvern, PA

There are several treatment technologies available for soils that have elevated levels of chromium. The technologies applicable to a particular chromium site depend on the clean-up goals, the form of the chromium present, and the volume and physical chemical properties of the chromium-containing soils. In many cases the clean-up goals are based on the Cr(VI) concentration in the soils. Therefore, most of the available treatment technologies consist of (1) removing the Cr(VI)-containing soils from the site, (2) immobilizing the chromium so that it will not leach after treatment under field conditions, or (3) reducing the Cr(VI) in the soils to the Cr(III) state. This article discusses the treatment technologies available for each of these remediation strategies, their advantages and disadvantages, and their relative treatment costs. Technologies evaluated include excavation and off-site disposal, soil washing, soil flushing, electrokinetics, solidification/stabilization, vitrification, and chemical and biological reduction. Particular emphasis is given to evaluating the technologies' ability to treat soils that have Cr(VI) throughout the soil matrix rather than simply on the surface of the soil particles.

1058-8337/97/$.50

C HROMIUM (Cr) is found in two oxidation states in the natural environment: the trivalent [Cr(III)] and the hexavalent [Cr(VI)] forms. Of these two forms, essentially immobile Cr(III) compounds are the predominant species in most environmental settings. Cr(III) is an essential dietary element for humans and is not classified as a carcinogen. Dietary supplements of 200 µg/d of Cr(III) are sometimes recommended for patients with adult-onset diabetes or insulin resistance (Fisher, 1990). There appears to be no evidence that naturally occurring background concentrations of Cr(III) in U.S. soils (ranging from 1 to 2000 mg/kg) pose any adverse health threat (Dragun and Chiasson, 1991; WHO, 1988). Region III of the U.S. Environmental Protection Agency (USEPA) has established a risk-based residential soil cleanup level for Cr(III) of 78,000 mg/kg based on an ingestion pathway (USEPA, 1996a). Likewise, recently the New Jersey Department of Environmental Protection (NJDEP) proposed a soil cleanup standard of 78,000 mg/kg for Cr(III) (NJDEP, 1995).

In contrast, Cr(VI) is highly mobile in the environment, is acutely toxic at moderate doses, and is classified as a known respiratory carcinogen in humans (USEPA, 1996b). Although the occurrence of Cr(VI) minerals is very rare in nature, hundreds of sites in the U.S. have significant levels of Cr(VI) in soils (HWC, 1993). The Cr(VI) is typically from industrial activities such as metal finishing and chemical manufacturing (Richard and Bourg, 1991). The risk-based residential soil cleanup guideline established by USEPA Region III for Cr(VI) based on an ingestion pathway is 390 mg/kg (USEPA, 1996a). This is more than 2 orders-of-magnitude less than the comparable Cr(III) guideline. USEPA has also designated a soil screening level of 270 mg/kg Cr(VI) for potential human exposure by inhalation (USEPA, 1996c) There is no comparable soil screening level for Cr(III).

AVAILABLE TREATMENT METHODS

There are a number of treatment technologies available for soils that have elevated levels of Cr(VI) other than conventional remediation by excavation and off-site disposal. The technologies applicable to a particular Cr site depend on the cleanup goals, the form(s) of Cr present, and the volume and physical/chemical properties of the Cr-containing soils. In many cases soil cleanup goals are based on the Cr(VI) concentration. This has been the case more recently since USEPA established distinctly different risk-based criteria for Cr(III) and Cr(VI) in soils. From a practical perspective, it is rare that Cr(III) concentrations in excess of 78,000 mg/kg are encountered except in soils subject to a Cr(III) chemical spill or natural soils high in chromite minerals. Because Cr(III) is naturally occurring and most forms found in the environment have low solubilities, and therefore low mobility in soils,

Cr(III) does not pose a significant groundwater or surface runoff hazard. This lack of risk was the rationale used by USEPA to make a no-action decision at a site where the soils had total Cr concentrations of almost 5000 mg/kg in the upper 6 ft of soil. The estimated volume of chromium-contaminated soil exceeding near background conditions (50 mg/kg of total chromium) was 1200 yd^3. The chromium present in the soil was predominately (95 to 99%) in the trivalent state. Another cleanup criterion often used for sites is the Toxicity Characteristic Leaching Procedure (TCLP) limit for total Cr, which is 5 mg/l (40 CFR 261.24). This procedure measures leachable Cr, which typically is predominantly Cr(VI). Wastes/soils that exhibit leachable Cr concentrations greater than 5 mg/l are defined as characteristic hazardous wastes and must be managed as such.

Consequently, most of the available remediation technologies for Cr(VI)-contaminated soils, apart from removing the Cr(VI)-containing soils from a site, consist of: (1) removing the Cr(VI) from the soils, (2) immobilizing the Cr so that it will not leach as defined by the TCLP test, or (3) irreversibly reducing the Cr(VI) in the soils to the Cr(III) valence state. Removal technologies include soil washing or flushing with and without chemical additives and/or the inducement of electrolytic forces. Immobilization technologies include solidification/stabilization, encapsulation, and vitrification. Reduction technologies include chemical and/or biological processes. Immobilization and reduction technologies may be performed either on excavated soils (*ex situ*) or on soils as they exist in the ground (*in situ*). This article discusses conventional and innovative treatment technologies and their applicability to soils that are contaminated with Cr(VI).

Removal Technologies

Excavation and Offsite Disposal.   A traditional remediation strategy consists of excavation and removal of Cr(VI)-contaminated media from a site and replacement with clean fill. However, this cannot be considered a treatment technology because the waste/soil characteristics are not changed by simply relocating the material. In fact, the land disposal regulations promulgated subsequent to enactment of the Hazardous and Solid Waste Amendments of 1984 to the Resource Conservation and Recovery Act (RCRA) of 1976 require waste/soils that do not meet the TCLP criteria for leachable Cr to be pretreated (e.g., reduction of Cr(VI) to Cr(III) and solidification) prior to land disposal (Marvin, 1993).

Excavation has an advantage over most technologies in that it is quick to implement and uses readily available equipment. It also removes the Cr(VI) in the excavated materials from a site, thereby allowing for negligible or low subsequent operation and maintenance (O&M) costs. The principal disadvantage of this approach (other than the relatively high cost of excavation, transportation, and disposal at a treatment, storage, and disposal [TSD] facility) is that it can cause Cr(VI), which was previously buried, to become airborne during excavation activities and therefore pose a threat to human health. As Cr(VI) is classified as an inhalation carcinogen by the USEPA, this should be a concern to anyone considering this approach.

Another disadvantage is that when Cr(VI)-bearing soils are identified beneath a building, removal or destruction of the existing structure may be necessary to gain access for remediating the underlying Cr(VI)-bearing soils. Excavation can also be complicated by the presence of Cr(VI)-contaminated soils below the water table. Removal of such soils from the saturated zone may necessitate the construction of retaining walls and a groundwater extraction system. Groundwater collected from the extraction system may need to be treated for Cr(VI) removal before discharging the treated groundwater back into the environment. Residuals containing Cr and other potentially toxic metals removed from the groundwater will need to be properly managed. Each such ancillary activity can significantly increase the overall remediation cost for a site.

Even without groundwater extraction and treatment, excavation and offsite disposal is costly, especially for deep Cr(VI)-contaminated soils. Although excavation itself may cost from $5 to $10 per ton, the treatment, transportation, and disposal costs typically exceed $100 per ton and sometimes $200 per ton, depending on the quantities and other characteristics of the waste/soils. Remedial investigation and implementation of a remedy designated as excavation and disposal at a permitted TSD facility of more than 100,000 tons of soils containing chromite ore-processing residue (COPR) from 32 small residential sites in Jersey City, NJ, was estimated to cost approximately $30 million (NJDEP, 1990). Because of these concerns, excavation and offsite disposal of Cr(VI)-contaminated soils is likely to be cost-effective only for sites with relatively small volumes of contaminated media and where the mobilization costs for other technologies would be expected to be a large portion of the overall site remediation cost.

Soil Washing. Soil washing, as it is discussed here, includes processes where excavated soil is mixed in a reactor with a washing solution to extract the contaminant(s) of concern from the soils. This includes washing with water or with other solvents. Other processes considered to be soil washing include metals extraction and heap leaching. Soil flushing, as discussed later, is the *in situ* application of soil washing.

The washing solution for a given application is selected based on the soil and contaminant chemistry. Chemicals are sometimes added to the washing solution to increase the solubility of the contaminants. Whereas surfactants are usually used with organic contaminants, chelating agents or acids are usually used to extract metals. Basic solutions may be needed for some metals, including some chromate compounds. Chelating agents, such as ethylenediamine tetraacetic acid (EDTA) and nitrilotriacetic acid (NTA), are compounds that form stable complexes with cationic metals. Because these complexes are more soluble in the washing solution than the uncomplexed metals, the washing solution typically removes more of the metals than it would without the chelating agent. Acids are typically used to remove inorganic compounds by converting them to soluble salts that dissolve in the washing solution.

In a recent U. S. Army Engineer Waterways Experiment Station bench-scale study, eight extracting agents (four acids and four chelants at three concentrations each) were applied to three different types of soils contaminated with cadmium, chromium, and lead at wide-ranging concentrations. Total Cr concentrations from 500 to 3000 mg/kg (almost all in the trivalent form) were poorly extracted by all the agents except fluorosilicic acid (HWC, 1996a).

Although these were indicative of soils found at military installations, Cr-contaminated soils at other sites more typically contain Cr as the chromate ion. Because the chromate ion is highly soluble in water, chelating or other complexing agents and/or acids are generally not applicable to a Cr(VI) soil-washing process. In fact, Cr(VI) is most soluble in alkaline media (e.g., pH > 11) (Vitale *et al.*, 1994). Because the solubility of Cr(VI) in aqueous solutions decreases with decreasing pH and its adsorptive capacity on soils increases with decreasing solubility, soil washing for Cr(VI) removal using an alkaline solution is likely to be more effective than water alone (James *et al.*, 1995).

Virtually all soil-washing or soil-flushing systems are designed to treat soils where the majority of the contaminants are concentrated in the finer-grained materials or on the surfaces of the larger soil particles. Many soil-washing processes are simply screening processes that separate the fine, contaminated particles from the bulk of the soil. The large particle fraction, which is typically the majority of the soil, is then clean and does not need further treatment before it can be placed back onsite. Other soil-washing processes combine screening processes with a scrubbing step that removes the outer layer of contaminants from the remaining larger particles. Several unit operations may be used in soil washing, including mesh sieves, mills, flotation units, and/or sedimentation basins. Where Cr(VI) is present on the surfaces of soil particles, such as when the soil is contaminated by a chemical spill, washing the soil surface with water and treating the resulting wastewater has proven to be feasible (McPhillips *et al.*, 1991).

Soil washing is not a destructive technology; the contaminants are only transferred from one contaminant phase to another. The resultant washing solution must then be treated to remove the contaminants and reduce the aqueous volume requiring disposal. Conventional technologies are available to treat soil-washing solutions and manage the residuals derived from them, but these technologies can significantly increase the cost for a remedial action. Several treatment options (chemical reduction and hydroxide precipitation, bioreduction, etc.) are discussed in subsequent sections of this article.

Once the washing solution has been treated, it is often recycled to treat subsequent batches of contaminated soil. The contaminated fine-grained materials that are separated from the washing solution also must be treated and/or disposed. The separation and treatment of these fines can be difficult. Clarifiers, polymer feed systems, filters, and/or other phase separation equipment are often needed to remove the fines from the washing solution and reduce their moisture content prior to disposal.

According to USEPA, typical metals removal efficiencies for soil washing range from 75 to 99% (USEPA, 1990; USEPA, 1992). The efficiency is dependent on a number of factors, including the length of time the soils have been exposed to the metals of concern, the amount of fines in the soil, and the affinity of the contaminants for the washing solution. Several authors state that if a soil has greater than 20 to 30% fines (with particle sizes less than 0.06 mm in diameter), then soil washing is probably not the most effective treatment technology available (Asp, 1990; Oravetz et al., 1992; USDOD, 1994). As noted by Oravetz, "soil washing for metal reduction is most likely effective only within a narrowly defined set of soil types, waste constituents, and operating parameters". For soils with the contaminants adsorbed to the larger particles the limiting factor is mass transfer.

Several references cited treatment of Cr-contaminated soils using soil washing without specific distinction between the Cr(III) and Cr(VI) soil concentrations. MTA Remedial Resources reportedly was able to remove 96% of the total Cr in a bench-scale test of its process on contaminated soils. In a pilot-scale test on contaminated soil from a wood-preserving site, the BioTrol Inc. soil washing system removed 47 to 83% of the total Cr. In bench-scale tests for USEPA's Superfund Innovative Technology Evaluation (SITE) Program, IT Corporation's Metal Extraction Treatment Process reduced the concentration of Cr from 244 mg/kg to 97.9 mg/kg, a reduction of 60%. Several USEPA Records of Decisions from wood-preserving sites specify soil washing or soil flushing for the removal of Cr (Palmetto Wood Preserving Site in South Carolina [1987]; South Cavalcade Street Site in Texas [1988]).

Bench-scale tests of a heap leaching process on soils that had Cr(VI) sprayed onto the surface of the soil particles showed promising results (Hanson et al., 1993). The heap leaching process consisted of pumping a leachate solution (tap water) into the top of a soil column, allowing the solution to dissolve the Cr(VI) from the surface of the soil particles, and then removing the Cr(VI)-containing leachate from the bottom of the soil column. Virtually all of the initial approximately 400 to 800 mg/kg of Cr(VI) was removed from the soil column after the application of only 1 pore volume of the leaching solution.

However, in some cases the Cr(VI) in the soils is an integral constituent of the soil or solid matrix itself. For example, more than 200 sites in New Jersey have been identified with suspected deposits of COPR generated from Cr chemicals production at three former facilities (James, 1996). These processing facilities produced Cr(VI) salts by roasting chromite ore with lime and soda ash in a kiln under oxidizing conditions (Shreve, 1967). This oxidized a large portion of the Cr(III) in the ore to Cr(VI). Much of the water-soluble Cr(VI) was then extracted from the roasted ore by leaching with water, followed by other steps to produce chromic acid, sodium dichromate, and other chromium compounds (Page and

Loar, 1991; Westbrook, 1991). The remaining COPR was highly alkaline and contained Cr(VI) due to incomplete leaching. From 1905 to 1976, the production of chromate salts in Hudson County, New Jersey, resulted in the production of approximately 2 million tons of COPR (NJDEP, 1990). Before the potential health hazards associated with Cr(VI) became widely known (Environmental Science and Engineering [ESe], 1989; Lioy *et al.*, 1992), this pebble-like residue was often mixed with other "clean fill" material and used to reclaim local low-lying and marshy areas, known as "the meadowlands". The COPR placed on the sites typically contained total Cr concentrations of 3 to 7% (ESe, 1989). Approximately 1 to 5% of this Cr was Cr(VI), with the remainder existing in the trivalent Cr(III) state (Sheehan *et al.*, 1991). However, as the COPR was mixed with other fill materials and has weathered over time, the Cr(VI) concentrations in the soils at these sites now vary widely.

Because the Cr(VI) in the COPR soils is an integral part of the media and not just a surface contaminant, it is very difficult to remove using soil washing. CH2M Hill performed bench-scale treatability tests to evaluate the feasibility of soil flushing to remove Cr(VI) from COPR-bearing soils. The COPR-containing soils were initially screened to remove particles greater than a 0.25 in diameter. In the test, a constant flow of deionized water was percolated downward through 10 in of the soils packed in a 3.25-in diameter column. Water samples were collected daily from the effluent of the column for pH and Cr(VI) analyses. After 36 d, over 200 pore volumes of water had passed through the column and the Cr(VI) concentration in the leachate had reached an asymptotic concentration of approximately 32 mg/l, after an initial peak of 350 mg/l. The column flushing operation was then terminated and soil samples were collected and analyzed for TCLP Cr. The TCLP Cr in a sample from the top of the column was less than the characteristic hazardous waste criterion of 5 mg/l. The four remaining soil samples collected from depths of 2.5 to 10 in within the column had TCLP Cr concentrations of 49.6 to 92.4 mg/l, thereby demonstrating that COPR will continue to release Cr(VI) from the interstices of the soil-like matrix much longer than typical soils.

Figure 1 presents the relationship between the Cr(VI) removed and the volume of water passed through the column during the test. These results indicate that although soil washing removed some of the Cr(VI) from the COPR-containing soils, large volumes of water were insufficient to render the treated soils non-hazardous based on the TCLP test. Therefore, unless the COPR-bearing soils were to be altered (ground or pulverized) to allow the deeply embedded Cr(VI) to be more accessible to a washing solution, the washing process would continue to be significantly limited by the slow diffusion of Cr(VI) from the interior to the exterior of the COPR particles. Even if the washing process would have been successful, the washing solutions generated by the process would need to be treated before discharge. This treatment would generate a residual sludge that would require appropriate treatment prior to disposal at a permitted TSD facility. The treatment of the washing solution and the residual sludge would be in addition to

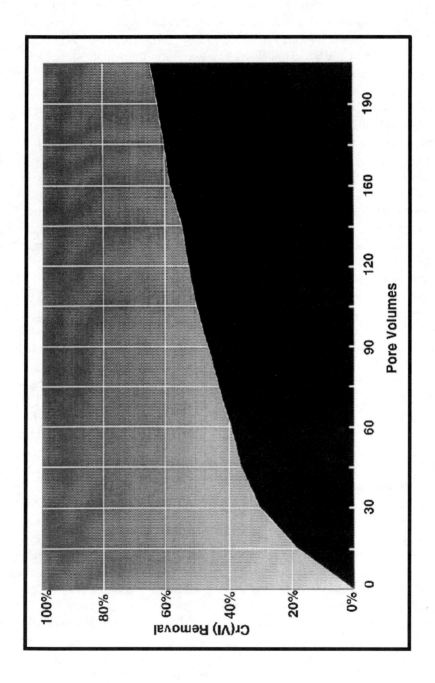

FIGURE 1

Chromium removal from COPR by flushing with water.

the soil-washing operation, thereby substantially increasing the overall cost for soil washing COPR-bearing soils.

Although it may not be applicable to COPR-bearing soils, soil washing is an innovative technology that has been used in commercial-scale applications to remediate soils contaminated with organic and inorganic contaminants, including more readily leachable forms of Cr, from more than a dozen Superfund sites. At least 15 vendors have full-scale systems and many of these full-scale units can process 50 to 150 tons of soil per day.

Remediation costs provided by several vendors for *ex situ* processes range from approximately $50 to $200 per ton. These costs include mobilization and demobilization, excavation, chemical additives to the washing solution, treatment of the contaminated soil, and treatment of the resulting contaminated washing solution, but typically do not include residuals management. The cost range assumes that the contaminant(s) are easily removed by the washing solution, the treated soils are redeposited on site, and the fines content does not exceed 20 to 25%.

Soil Flushing. *In situ* soil washing is often referred to as soil flushing. The washing solution may be delivered to the soil through a series of injection wells or it may be applied to the surface of the soil and allowed to percolate downward through the contaminated media. The washing solution is then recovered through recovery wells or interception galleries in trenches. Soil flushing can be complicated by a variety of chemical reactions that can take place in the soil. Armienta and Quéré (1995) found significant adsorption of Cr(VI) by silt and clay layers in a Cr(VI)-contaminated area in Mexico, making it difficult to restore the groundwater quality to precontamination conditions using conventional pump and treat remediation. As with soil washing, soil flushing requires that the Cr(VI) be on the surface of the soil, and therefore would not be feasible for sites where the Cr(VI) is deeply embedded within the soil particles.

Soil flushing has been used to remediate chromium sites. The United Chrome Superfund Site in Corvallis, Oregon, had Cr(VI) concentrations in the site soils greater than 60,000 mg/kg and in the groundwater at greater than 5000 mg/l (McPhillips *et al.*, 1991). The source of the Cr(VI) was leakage from plating tanks and rinse water discharges from a hard-chrome plating shop. The remedial action for the site consisted of excavation and offsite disposal of approximately 1100 tons of highly contaminated soil; installation of groundwater extraction and monitoring wells; and construction of two percolation basins and one infiltration trench. By flushing the Cr-contaminated soils with water from the infiltration units and then extracting the Cr-contaminated water through the extraction wells, USEPA was able to remove 24,300 lb of Cr(VI) from the site's soils and groundwater. This Cr(VI) was contained in 6.7 million gallons of extracted water, which were then treated on site in a chemical reduction and precipitation system. This volume of water represented the extraction of an average of only 1.5 pore volumes to reduce the Cr(VI) concentration to approximately 40 to 50 mg/l. However, the field data indicated that it may not be possible to meet the groundwater cleanup criterion of 10 mg/l at one of the well locations due to the presence of solid-phase Cr(VI).

Because the effectiveness of soil flushing for metals removal is so dependent on the physical/chemical properties of the soil particles to be treated, the cost of applying this technology is as variable as the soil conditions that may be encountered. The principal difference between applying soil flushing to Cr(VI)-contaminated soils compared with Cr(III)-contaminated soils is the cost for the chemical agents needed specifically for the removal of Cr(VI) vs. the chemical additives required to remove trivalent forms of Cr.

In general, the cost range for removing Cr(VI) from soil (excluding those containing COPR) is expected to be similar to that for other metals. Recognizing that soil flushing has been used infrequently for metals removal at Superfund and other sites due to performance limitations, the unit costs for soil flushing are projected to range from about $75 to $200 per ton of soils treated (HWC, 1996a).

An innovative technology that is being developed for use with soil flushing is electrokinetics. This is a process that uses an electric current to move ionic contaminants through a soil to a collection well. The contaminants are then removed from the well for further treatment or disposal. In electrokinetics, an electric current is applied to water-saturated soil by an array of anodes and cathodes, as illustrated in Figure 2. The application of this electric current causes three potentially beneficial phenomena to occur in the soil-water system: electroosmosis, electrolysis, and electrophoresis.

Electroosmosis causes the moisture in saturated soils to migrate from the anode to the cathode in low permeability soils. Electroosmosis has often been used to increase the natural hydraulic gradient in the dewatering of a fine grained soil for construction. It can also be used to increase the flow of cations (positive ions) in contaminated groundwater toward a recovery well located at the cathode (USEPA, 1989a).

Electrolysis causes the anions (negative ions) and cations in the groundwater to migrate toward the anode and cathode, respectively. As the ions build up at the electrodes, this migration decreases due to the interaction between the similarly charged ions. In tests of the electrokinetics process for remediating contaminated sites, groundwater containing cationic contaminants has usually been withdrawn at the cathode to prevent the buildup of cations and enhance the effects of electrolysis (Cabrera-Guzman et al., 1990). Two of the major ions that can impact the concentration of ions at the electrodes are the hydrogen (H+) and hydroxyl (OH–) ions. These ions are produced by the dissociation of water at the electrodes. The formation of these ions at the electrodes can cause the groundwater to become basic near the cathode and acidic near the anode (USEPA, 1990). Over time, an acid front may migrate from the anode toward the cathode. This migration may cause desorption of contaminants from the soil.

Electrophoresis also causes the charged soil particles to migrate to the electrodes. In electrophoresis the charged particles move through a stationary liquid.

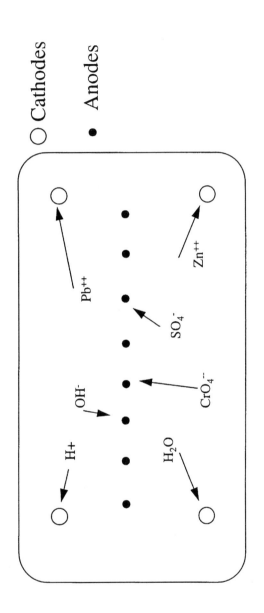

FIGURE 2

Sample electrokinetics diagram.

777

In most applications of the electrokinetics process the effect of this phenomenon is minimal (Cabrera-Guzman *et al.*, 1990). It could be significant if the soil consists of small, charged particles. This could lead to these small particles plugging the groundwater migration pathways and interference with the effects of electroosmosis and electrolysis.

Electrokinetics will affect the two principal forms of Cr differently because Cr(VI) is generally a chromate or dichromate anion and Cr(III) is essentially always a cation in environmental media. Soluble Cr(III) migrates toward the cathode. This is expected to be minimal in typical soils because Cr(III) has a low solubility at neutral pH. This migration would be expected to be enhanced by the effects of electroosmosis. Cr(VI) migrates toward the anode, counter to the electroosmosis effects.

Bench- and field-scale treatability tests of the electrokinetics process were conducted on Cr(VI)-contaminated soils from the United Chrome Superfund site (Banerjee *et al.*, 1987). These soils were contaminated by chromic acid from a chrome-plating facility. The field tests used one cathode well surrounded by an array of anodes. The distance between the cathode and the anodes ranged from less than 1 to 2.5 meters. The wells were 15 ft deep and contained steel reinforcing rods that were used as electrodes.

The results of these tests showed that Cr(VI) can be removed from contaminated soil by the application of electrokinetics and groundwater withdrawal at the anode. The results indicated that the most efficient treatment process used periodic rather than continuous groundwater withdrawal (due to slow rate of Cr release from the soil particles) and that some of the chromate was reduced to Cr(III) by the ferrous iron produced at the iron electrode. The tests also found that the effects of electrolysis can be best utilized to remove charged contaminants when the contaminant "has a low valence, high diffusivity, and relatively high concentration compared to other ions in the system" (Banerjee *et al.,* 1987).

For electrokinetics to be most effective, the soil to be treated should be saturated. However, as groundwater rarely saturates soil to grade elevation, supplemental moisture can be added to some unsaturated soil types (e.g., clays) to get the process to function. One way of increasing the soil moisture content would be water injection into wells at the cathodes of the electrokinetics system. This would not, however, guarantee that the contaminant (e.g., Cr) present within the unsaturated soils would be in contact with water because of potential channelling during percolation of water from the surface and uneven distribution of moisture inherent to the soil properties. Thus, the diffusion of moisture into the soils and the extraction of Cr may be a slow process. Additionally, if the Cr is integral to the molecular structure of the particles, as with COPR, or is highly sorbed into their interstices (e.g., aged contamination), electrokinetics processes can be expected to be less effective than for soils in which the Cr(VI) can be readily dissolved in the pore water.

Another potential use for electrokinetics in Cr(VI) remediation is the distribution of treatment chemicals. If an ionic reducing agent is applied to the soils at a site, electrokinetics could be used to speed up the distribution of the chemicals within the soils and simultaneously reduce the Cr(VI) present at the anodes.

Electrokinetics works best in clayey soils with low permeabilities. Clay has a high ionic exchange capacity, which enhances the electroosmosis effect, and a low permeability, which makes the effects of the hydraulic flow less than those of the ionic flow. In other words, in soils with a high permeability, the soil washing effects from groundwater withdrawal at the electrodes would be expected to mask the smaller beneficial effects from the actual electrokinetics process.

The application of electrokinetic processes to Cr-contaminated soils in bench-scale testing has shown reductions in total Cr ranging from 75 to 95% from initial total Cr concentrations up to 1000 mg/kg in kaolinite clay soil. As such testing was performed only for comparative effectiveness, no projections were prepared for full-scale application (Banerjee *et al.,* 1987).

Remediation costs for metals removal in general are projected to be in the range of \$20 to \$150/yd^3. These estimates are based on bench- and limited pilot-scale field testing, because this technology is still in the developmental stage and field-scale applications are currently under evaluation. The energy demands of the electrokinetics process for an easily treated soil are estimated to be 20 to 60 kWh/m^3 of soil remediated. Using a power cost of \$0.10/kWh, this translates to \$1.50 to \$4.50/yd^3 of soil remediated. This cost is most likely minor compared with the equipment and associated groundwater treatment costs that would be required for the extracted groundwater.

Immobilization Technologies

Cr(VI)-contaminated soils can be treated *in situ* or *ex situ* to immobilize the Cr(VI) so that the treated soils will pass the TCLP test. *In situ* treatment has the advantage of minimizing the exposure of site workers and local residents to airborne Cr(VI). It also has the potential for minimizing disruption to or demolition of existing structures.

Solidification/Stabilization. The purpose of solidification and stabilization is to treat contaminated soils so that the contaminants are suitably immobilized from potential leaching into the environment. Solidification is the binding of a waste/soil into a solid mass to reduce its contaminant leaching potential, whereas stabilization is the reduction of the solubility and/or chemical reactivity of a waste/soil. Reduction is a specific stabilization technique that is discussed in subsequent sections of this article. These technologies are applicable to a wide range of wastes/soils, but are particularly well suited for metals and are typically limited to soils containing less than 1% organics (HWC, 1996a). Solidification/stabilization (S/S) can be done *in situ* or *ex situ* on excavated materials by processing at a staging area either on site or off site.

The waste/soil is solidified by the addition of admixtures, which can include Portland cement, fly ash, lime, cement kiln dust, and/or other pozzolanic materials and water, if necessary (Conner, 1990). The admixture addition is followed by mechanical mixing to distribute the admixture throughout the matrix. The admixture reacts chemically with metals in the soil, typically to form complex insoluble silicates and hydrates. During these reactions, excess acidity is also consumed. Over a period of approximately 28 d, the waste/soil matrix solidifies and cures, yielding a concrete-like final product.

S/S has several advantages relative to other hazardous waste treatment technologies. The process is relatively inexpensive, with the majority of costs being in the procurement of cementing agents. It has a long, demonstrated performance record for radioactive and mixed waste treatment and as a pretreatment method to meet land disposal restrictions for hazardous wastes in accordance with Resource Conservation and Recovery Act (RCRA) regulations. As a result, there are more performance data available for S/S than for most other soils treatment technologies.

A conventional variation of the S/S process is to first reduce the Cr(VI) to Cr(III) and then encapsulate the soils using polymeric or asphaltic materials. These materials effectively contain or seal the metal contaminants in the soil matrix to minimize potential leaching due to contact with environmental moisture. In certain applications, the polymeric or asphaltic material is applied at elevated temperatures to enhance its distribution and binding tenacity. Although encapsulation has been used routinely for radioactive waste treatment, it is typically not competitive with other S/S processes that use less-expensive stabilizing and solidifying agents for metallic-contaminated soils.

A major concern for applying S/S to Cr(VI)-bearing soils is that the Cr(VI) must be reduced to the trivalent form prior to solidification. The waste/soil would not be effectively stabilized if Cr(VI) were present, due to its high mobility, even in a concrete matrix. This adds a layer of cost and technical complexity for remediation of Cr(VI)-contaminated sites compared with sites with other metal contaminants. As noted by Conner (1994), reduction of Cr(VI) in wastewater is a straightforward process, whereas reduction of Cr(VI) in a solid matrix is not.

It is important to recognize, however, that the S/S process does not remove Cr from the treated waste/soil. Therefore, the process is not a viable strategy where total Cr levels are in excess of regulatory limits, although the current USEPA guideline for Cr(III) in soils (78,000 mg/kg) is greater than all but rare cases of Cr(III)-contaminated soils. The application of S/S, in conjunction with chemical reduction, can reduce the mobility of Cr(VI) in waste/soils sufficiently to meet the TCLP criterion for Cr. Thus, the S/S technology can be applied effectively at locations where total Cr concentrations are within acceptable limits, but only after sufficient testing has been performed to determine performance effectiveness and overall process application costs.

Sites with a high water table present a considerable engineering challenge and process constraint to using S/S processes. One S/S process configuration that can be used at these sites that does not require sheet piling and groundwater removal is *in situ* mixing with a hollow-stem auger. This process is viable as long as sufficient soil bearing capacity is available to support the equipment. Several companies, including Geo-Con, Inc., Millgard Environmental Corporation, and Chemical Waste Management, Inc., have this capability. The hollow-stem auger can deliver reducing admixture(s) and mix the waste/soils *in situ* to depths up to 100 ft. This reduces the amount of equipment needed to isolate or otherwise manage shallow groundwater.

Geo-Con estimated that the remediation cost of a Florida site contaminated with PCBs and lead with a high water table using a rig with one auger was approximately $190 per ton (USEPA, 1989b). The drilling/injection reached a maximum depth of 18 ft in treating about 300 yd^3 of soil with a 3-ft-diameter auger. It was projected that the cost for treatment could be reduced to about $111 per ton using a rig with four augers for larger applications (Stinson, 1990). Two principal factors affecting the project costs are the type of admixture used and the proximity to an admixture source. The costs of admixture can vary significantly depending on the material selected (e.g., Portland cement is more expensive than fly ash) and the local availability of the material. The costs for the drilling and mixing operations alone are estimated to be between $10 and $30 per ton. These costs are dependent on the size of the site, depth of contamination, the degree of difficulty and rate of mixing the admixture and waste/soils, and the degree of dewatering required. The costs for conventional *ex situ* solidification/stabilization processes range from $40 to $110 per ton, including excavation (USEPA, 1986; Arniella and Blythe, 1990; USDOD, 1994).

The application of S/S technology has the potential to achieve site cleanup in a relatively short period relative to some of the other technologies discussed. This would result in a large short-term cost. Other technologies that require much longer periods of time but achieve similar cleanup limits could cost much less on an annual basis. Thus, it is important to determine the unit cost (e.g., per ton or yd^3) on a present worth basis to compare different technologies that have similar performance characteristics.

Bench-, pilot-, and full-scale testing should be performed to screen and select admixture materials, set mix ratios, and verify acceptable performance before selecting this technology. It is also important to evaluate dust and vapor emissions, which can pose a significant problem during construction, when Cr(VI)-contaminated soils are the object of treatment. Specifically, tests should be conducted to assess the silica content of the waste/soil, selective physical and chemical properties, and the Cr TCLP of the treated soils. A high silica content, for example, would favor the S/S process. Additionally, the nature and concentrations of anions should be characterized to evaluate potential interferences. Long-term stability (i.e., by freeze-thaw testing) of the solidified product is another important consideration.

Vitrification. Hazardous wastes can be incorporated into a solid matrix by the process of vitrification. The waste matrix (contaminants and soil) is heated to a molten state and allowed to cool, forming a material similar to volcanic glass with compressive and tensile strength approximately 10 times greater than unreinforced concrete. Vitrification can be done *in situ* or after excavation in an external reactor. Organic contaminants are volatilized or oxidized at process temperatures typically exceeding 2000°F. Metals are entombed in the chemically and physically inert product resembling obsidian. During this process it has been reported that Cr(VI) can be reduced to the Cr(III) state (Stanek, 1977). Vitrification of a waste matrix can reduce the waste/soil's toxicity, mobility, and volume, and yields a product that can be demonstrated to be non-hazardous based on current hazardous waste test criteria. However, vitrification is seldom applied for heavy metals immobilization due to the high energy requirements and much greater overall unit cost compared with other technologies.

In situ vitrification (ISV) has been marketed by Geosafe Corporation, a subsidiary of Battelle Pacific Northwest Laboratories, which developed the process for application at sites with mixed wastes (radioactive and chemical contamination). The Geosafe process incorporates electric melting of the waste matrix at 1600 to 2000°C to form a glass and crystalline product. Four electrodes are embedded into the ground, typically in an 18-ft square (Figure 3). Glass frit and graphite are placed on the soil surface to provide a conductive path between the electrodes. Electric current (up to 4000 V) is passed between the electrodes to melt the soil to the desired depth. Offgases are drawn through a hood to an offgas treatment train. This configuration can produce a melt 30-ft by 30-ft by 19-ft deep. Additional configurations can be activated simultaneously or sequentially to treat larger volumes. After the melt cools (up to 1 year), the vitrified monolith can be left in place.

ISV should generally be limited to soils greater than 6 ft and less than 20 ft below the surface that contain less than 7% combustible liquids and no more than 3200 mg/kg of combustible solids (HWC, 1996a). Its application may be complicated by a high water table or elevated soil salinity. Barriers such as slurry walls may be required to prevent the inflow of groundwater. Additionally, control over the depth of melt to preclude damaging an existing naturally occurring Cr(VI) barrier at a site, such as the previously mentioned meadow mat, is not possible with existing ISV technology.

The ISV process has been demonstrated to be effective in the treatment of waste/ soil matrices containing heavy metals, liquid organics (including PCBs), solid organics, and radioactive materials. The process destroys hazardous organics and the vitrified product typically meets all TCLP limits for metals and organics. ISV can be expected to reduce the volume of the waste/soil matrix by 20 to 50%. Treatment rates currently range from 26,000 to 44,000 tons per year. The U.S. Department of Defense (USDOD) (1994) listed average mobilization/ demobilization costs for the ISV equipment in the range of $200,000 to $300,000. The operational costs of the system vary with electricity costs, the amount and cost of

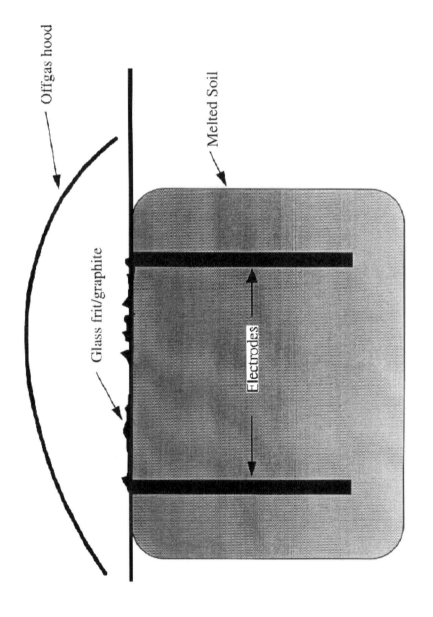

FIGURE 3

Sample in situ *vitrification diagram.*

783

any required additives, the level of groundwater, and the depth of the contamination. Geosafe used the results from a demonstration project for the SITE program to estimate treatment costs of "\$740/ton, \$430/ton, and \$370/ton for 5-ft, 15-ft, and 20-ft deep treatment cells, respectively" (HWC, 1996b). If other less costly technologies cannot be applied effectively and the limitations of ISV are not encumbering at a site, appropriate bench- and pilot-scale tests are needed prior to the selection of ISV as a remediation technology.

Several *ex situ* thermal treatment processes have been developed as part of the SITE Program. These include plasma centrifugal furnace (PCF), cyclone furnaces, electric melting, and oxidation/vitrification. These processes all incorporate a precombustion/feed step (usually proprietary), high-temperature melting, and cooling to form the vitrified product. The products are also reported to meet all TCLP requirements, and their application to contaminated soils is considered a final solution for the treated hazardous waste/soil. The fundamental differences among the various *ex situ* technologies are in the electrical requirements and the operation and maintenance costs.

The Plasma Arc Centrifugal Treatment (PACT) process marketed by Retech, Inc., a division of M4 Environmental Management, Inc., uses a plasma torch to create a molten bath that detoxifies the feed material and vitrifies it between 1500 and 1600°C. The process has been tested at numerous sites, including a Department of Energy test site in Butte, Montana, a commercial site in Switzerland, and a commercial facility in France. The process has not yet been applied to a hazardous waste site. Retech estimates that a full-scale system could process between 8500 and 22,000 tons of contaminated waste/soils per year.

The Vortec Corporation Combustion and Melting System (CMS) is another *ex situ* process developed with support from the SITE Emerging Technology Program. The CMS furnace can be fired with a variety of fuel sources, including coal slurries, natural gas, and waste fuels. The unit's capacity may reach 140,000 tons per year.

Vitrification is expected to reduce the mobility of Cr significantly. Preliminary results indicate that Cr(VI) can be reduced successfully to Cr(III) and incorporated into the vitrified product. However, Cr is not physically removed from the waste. Thus, sites that contain total Cr concentrations above regulatory limits cannot be adequately remediated using this technology alone, unless the site remedy includes a provision for acceptance of the treated waste/soil based on Cr leachability characteristics.

Major technical questions remain on the full-scale operation of both *in situ* and *ex situ* vitrification technologies for Cr(VI)-contaminated soils. At present, only pilot-scale and/or limited full-scale work has been done with most of the processes described here for treating hazardous wastes. Although many of the technologies have been used for years in conventional glass-making processes, hazardous waste treatment has been limited primarily to synthetic and radioactive wastes at federal energy facilities. The *ex situ* processes are at present unavailable for mobilization, and thus only ISV may be available for a full-scale remediation.

ISV requires approximately 750 kWh of electricity per ton of treated material, including any additives. The cost of *in situ* vitrification generally ranges between $350 and $400 per ton of treated waste. A large portion of projected operating costs is attributed to the amount of energy required to melt the soils. These estimates do not include site preparation activities, such as the construction of slurry walls, groundwater pumping and treatment, and pretreatment to reduce salinity (if necessary). The unit cost of *ex situ* processing may range from $90 to $700 per ton, depending on the scale of application, the availability of mobile equipment, possible pretreatment requirements, the fuel used, and/or the power costs. The lower part of the range also assumes that the resulting vitrified material can be sold as a product. Although unproven with Cr-contaminated wastes, many of the vendors contacted believe that there could be a market (yet to be demonstrated) for many vitrified materials.

The cost estimates for vitrification presented in this review are based on pilot-scale studies, limited full-scale application of a few vendor systems to waste/soils containing radioactive, organic, and/or mixtures with metallic contaminants, and theoretical scale-up models from such studies. These are only order-of-magnitude estimates that may be significantly different than the actual costs of treating Cr(VI)-contaminated soils at a particular site. As with most technologies, the unit cost for applying vitrification processes is expected to be inversely proportional to the volume of waste/soil to be treated.

While vitrification technology may hold promise for addressing sites with Cr(VI) contamination, the various processes are in the emergent stage for full-scale use in the field. Likewise, vitrification is significantly more costly than alternative technologies for most sites where metals-contaminated soil is the principal or only concern. Commercial *in situ* or *ex situ* vitrification process equipment would most likely require a year or more to fabricate for a given application and full-scale pilot testing would be required prior to implementation. These tests would be in addition to the high unit cost of applying this technology.

Reduction Technologies

As with immobilization, Cr(VI)-contaminated soils can be treated *in situ* or *ex situ* to reduce the Cr(VI) to Cr(III) and thereby reduce the toxicity and mobility of the Cr.

Chemical Reduction. Chemical reduction can be used to convert Cr(VI) to the trivalent valence state, which is generally less toxic and less soluble than most Cr(VI) compounds encountered in the environment. The chemical reduction of Cr(VI) in wastewaters is a well-established process that has been used in industry for decades (Patterson, 1985). Chromium plating and metal treating facilities and many other industries generate Cr(VI)-containing aqueous wastes and treat them by chemical reduction. The treatment of Cr(VI) in soils is sometimes more difficult as "various side reactions can occur that waste reagent and/or cause other problems. . . " (Conner, 1994).

In the reduction of Cr(VI), the Cr molecule accepts three electrons from an electron donor, or reducing agent, changing from the +VI oxidation state to the +III state. Industrial Cr(VI) reduction processes usually involve mixing the Cr(VI)-bearing wastewater with a reducing agent and precipitating trivalent Cr hydroxide [$Cr(OH)_3$].

There are many reducing agents commonly used in industrial Cr(VI) reduction processes, including

- Ferrous sulfate

- Ferrous ammonium sulfate

- Sodium sulfite

- Sodium hydrosulfite

- Sodium bisulfite

- Sodium metabisulfite

- Sulfur dioxide

Ferrous iron (FeII) reduces Cr(VI) at neutral and alkaline pH. Reduction of Cr(VI) by sulfite, bisulfite, metabisulfite, and sulfur dioxide occurs optimally at a pH of 2 to 2.5. Therefore, acid is usually added to decrease the pH of the wastewaters. After the Cr(VI) is reduced, the pH of the wastewater is increased to precipitate $Cr(OH)_3$.

Reducing agents can be delivered to the subsurface at hazardous waste sites by several methods, including injection wells, infiltration galleries, or *in situ* soil mixing equipment. Injection wells and infiltration galleries are often used in conjunction with groundwater extraction wells as part of pump-and-treat systems. They are likely to be of limited use for introducing reducing agents for chromium, however. Sulfur-containing compounds require a low pH, which cannot be attained in a normal soil matrix, due to degradation of carbonate minerals. Ferrous sulfate will work at high pH's but its low solubility above a pH of 9 limits the ability to deliver it to alkaline soils by infiltration or injection. *In situ* soil mixing is a relatively new remediation process in which high-torque augers and/or other equipment are used to mechanically mix chemical agents into the soil without excavation from the ground. *In situ* soil mixing is specifically intended to homogenize the soil to promote uniform distribution of the remediation chemicals, whereas injection wells and infiltration galleries rely on dispersion through the soils from the point of reducing agent application.

As discussed previously, *in situ* reduction is often performed in conjunction with *in situ* S/S to stabilize the contaminated material. *In situ* soil mixing can be used to mix solidification agents, such as Portland cement, into the Cr(VI)-contaminated material. Other agents, such as acids or bases, may also need to be mixed into the soil matrix if the soil does not have a suitable pH for the reduction reaction. For example, the COPR sites discussed previously typically have soils with pH values above 11. Because ferrous sulfate is not soluble in this pH range, it would be ineffectual to apply ferrous sulfate by injection or percolation at these sites. The material could be applied, however, using soil mixing techniques. In this case, the material could be distributed as a solid and it then would be available when needed for reducing the concentration of chromium dissolved in the groundwater. Even given its low solubility, a portion of the ferrous iron would be in solution and would react, allowing more iron to dissolve. Therefore, the problem with ferrous sulfate solubility is a delivery problem.

Cr(VI)-contaminated soil can also be remediated *ex situ* by excavating the soil, mixing it with a reducing agent or a combination of stabilizing chemicals, and backfilling the soil on site. As with other *ex situ* Cr(VI) remediation processes, this approach has the disadvantage that airborne particulates with Cr(VI), a known human health hazard, may be generated during remediation. Thus, more stringent health and safety procedures would likely be needed for *ex situ* treatment than for an *in situ* application.

The effectiveness of applying reducing agents to subsurface soils may be limited by the degree and rate of reducing agent penetration to the interstices of the Cr(VI)-contaminated soil and the tenacity by which the Cr(VI) is sorbed to the soil particles. The Cr(VI) inside the particles may leach only gradually over time, which could extend the time required to achieve the desired remediation goal. In some situations, an excess reducing agent may have to be applied initially or in sequential doses to address such a phenomenon. The rate of application of the reducing agent should be based on whether the reducing agent will be rapidly oxidized or flushed from the soils by other processes.

In situ soil mixing is intended to break up the soil mass and allow thorough mixing of the reducing agent with the soil particles. The reducing agent is typically delivered simultaneously by the mixing device. Several vendors were contacted concerning *in situ* reduction of Cr(VI)-contaminated wastes. They reported extensive laboratory-, pilot-, and commercial-scale experience with the *in situ* injection and mixing technology for S/S of metallic contaminants in general, but limited demonstration- and full-scale applications specifically for Cr(VI)-bearing soils.

In situ Cr(VI) reduction has been accomplished at several sites by injection of reducing agents into the subsurface. For example, it has been employed in a Superfund demonstration project at the Townsend Sawchain Facility site in Columbia, South Carolina, by Secor International Corp. At this site total chromium concentrations approach 10,000 mg/kg in the soils and act as a source of Cr to the groundwater. The groundwater in the shallow sandy aquifer contains Cr(VI) concentrations of 5 to 10 mg/l. For the demonstration, a ferrous sulfate solution was pumped into the subsurface using injection wells. The pH of the solution was adjusted to approximately 3 to keep the ferrous sulfate in solution. A full-scale remediation using this process is to begin in 1997. Secor also has used *in situ* reduction for the remediation of Cr(VI)-containing soils and groundwater for a confidential client at a chromium site in California, where the selected remedial action has been initiated. In less than 2 months, approximately 2 acres of Cr(VI)-contaminated soils have been remediated. The initial concentrations of Cr(VI) in the soils were as great as 200 mg/kg. The injection process has decreased these concentrations by 3 orders of magnitude to less than the Cr (VI) clean-up level of 200 ug/kg. Remediation of the groundwater is also on-going.

Geo-Con, Inc., Millgard Environmental Corporation, and *In Situ* Fixation Inc. all offer *in situ* soil mixing services for site remediation. Although they have not reported full-scale remediation experience with Cr(VI)-contaminated sites, both companies have successfully remediated Cr(VI)-contaminated soil by reduction and solidification in laboratory treatability studies.

A small site with soil contaminated by a Cr(VI) solution has been remediated *ex situ* in California. A representative of the California Department of Health Services who oversaw the site remediation stated that approximately 150 yd^3 of soil were excavated and treated by adding and mixing a reducing agent in an above-ground container. The treated soil was transported by truck to a TSD facility for disposal.

Vendor representatives have indicated that the unit cost of using *in situ* soil mixing equipment (excluding chemicals) for *in situ* reduction ranges from about $20 to $35 per cubic yard. The cost of chemical addition can vary widely, depending on the Cr(VI) level in the soil and other characteristics which may affect the oxidation/reduction (redox) status of the soils. The estimated cost for *ex situ* reduction ranges from about $75 to $100 per cubic yard (including chemicals and excavation). All of these costs can also vary depending on groundwater conditions, the possible need for solidification chemicals, and the volume of soil to be treated. For sites with relatively small volumes of Cr(VI)-contaminated soil (e.g., 200 to 300 yd^3), excavation and off site treatment and disposal is generally more economical than mobilizing the equipment for an on site treatment process.

Biological Reduction. Numerous researchers have demonstrated the ability of a variety of microorganisms to reduce Cr(VI) to the Cr(III) state in laboratory tests (DeFilippi, 1992; Ohtake and Hardoyo, 1992; Blake *et al.*, 1993; Mehlhorn *et al.*, 1994; Shen and Wang, 1994). Similarly, at the COPR sites mentioned earlier, it has been observed that the meadow mat, an organic-rich layer of decayed vegetation that underlies the Cr(VI)-bearing fill material, acts as a natural barrier to the vertical migration of Cr(VI). The groundwater in a silty sand layer directly beneath the meadow mat has been shown to contain nondetectable levels of Cr(VI). The meadow mat is under highly reducing conditions due to bacterial activity associated with the decaying organic material (James, 1996). Based on the observed ability of the meadow mat to reduce Cr(VI) to Cr(III), bench- and pilot-scale investigations were conducted to determine the potential for bioremediation at sites with COPR-enriched fill.

Laboratory column testing performed by the authors involved adding and mixing a mineral acid (to decrease the pH from >11 to between 6.5 and 9.5) and a bacteria-rich organic substrate (fresh cow manure) to the COPR-enriched soils (see Figure 4). The soils in the columns were saturated, capped, and left undisturbed except for periodic sampling events. The experimental results demonstrated effective bioreduction of the Cr(VI) to Cr(III). As shown in Figure 5, the pore water Cr(VI) was reduced from greater than 800 mg/l to less than 0.5 mg/l over a period of 3 months and to less than 0.1 mg/l over 7 months. Likewise, Figure 6 shows that solid phase Cr(VI) concentrations decreased from approximately 2000 mg/kg to less than 10 mg/kg in the columns in less than 11 months. The TCLP extracts from the treated columns met the regulatory limit (for characteristic hazardous waste) of 5 mg/l of Cr, whereas the untreated samples had TCLP extract concentrations greater than 45 mg/l. This study demonstrated the potential applicability of *in situ* bioreduction to soils contaminated with Cr(VI) by adjusting the pH to between 6.5 and 9.5 and admixing a bacteria-rich organic substrate. Subsequent field pilot testing of this process has been initiated, and preliminary results confirmed the potential application of this process to COPR-enriched soils, but mass transfer limitations associated with the gas formation and foaming during *in situ* injection/mixing of the mineral acid remain to be resolved.

Bioremediation processes have been used in full-scale projects to successfully remediate soils contaminated with petroleum products, wood preservatives, sol-

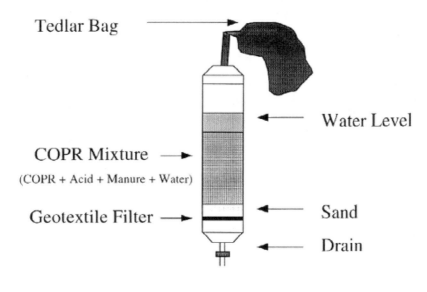

FIGURE 4

Bioreduction column aparatus.

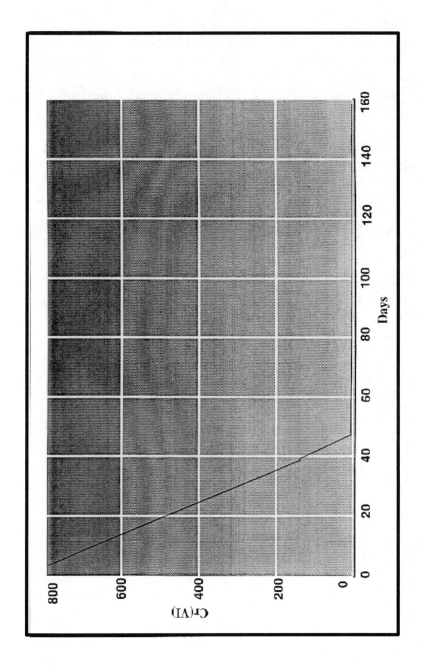

FIGURE 5

Liquid phase Cr(VI) bioreduction.

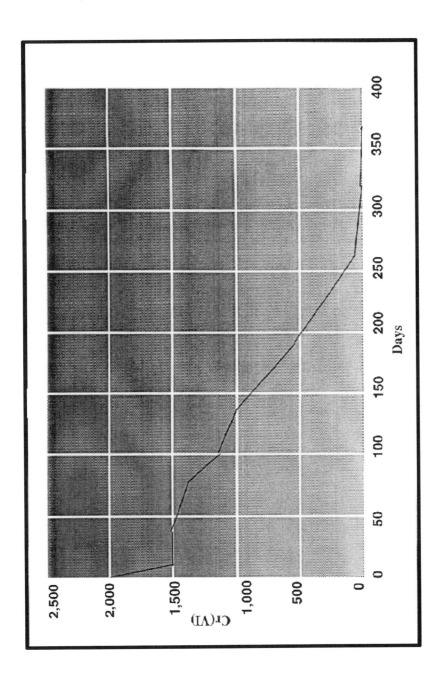

FIGURE 6

Solid phase Cr(VI) bioreduction.

vents such as alcohols, and other organic compounds. The typical unit costs quoted by vendors of the technology ranged from $20 to $100 per ton, with *ex situ* processes tending to be more costly than *in situ* processes due to excavation and materials-handling requirements. The *in situ* remediation of a petroleum-contaminated site at a U.S. Naval Communication Station in Scotland cost approximately $30 per ton (USDOD, 1994). In this application, bacteria and nutrients were applied using injection wells and surface sprayers. The unit cost for a Cr(VI)-contaminated site is expected to be comparable for the equipment-related elements, but more costly operationally due to the possible need to control the pH of the Cr(VI)-contaminated soil within an acceptable range for bacterial propagation. Prior to the selection of bioreduction as a treatment technology, bench- and pilot-scale tests should be conducted to determine the proper environmental conditions (pH, temperature, moisture content, etc.), and the organic and nutrient additive levels required for the sustained growth of the Cr(VI)-reducing microorganisms.

The advantages of an *in situ* bioreduction process are the same as for the other *in situ* processes (reduced worker exposure to airborne Cr(VI), lower materials-handling costs, etc.). In addition, the organic materials that are used in the process may have a greater long-term stability than the chemicals used in a rapid chemical reduction, although the use of excess chemical reductant remains a potential option for the latter reduction mechanism. This could be important for sites that contain Cr(VI) in matrices similar to COPR. As the Cr(VI) continues to leach (at decreasing rate) out of the soil matrix over time, there needs to be sufficient reducing agent available to react with the Cr(VI) and reduce it to Cr(III). In chemical reduction processes, the reducing agents are oftentimes consumed by other chemical reactions (such as with dissolved oxygen) or are washed out of the soil matrix before the Cr(VI) has leached sufficiently from the interior of the soil particles to the exterior and been reacted with the reducing agent(s). In biological reduction processes, however, the microorganisms appear to remain viable as long as sufficient organic substance and nutrients are available and the pH of the medium remains in a suitable range for propagation.

SUMMARY

Apart from removing soils for offsite processing and disposal, the treatment of soils for reduction of Cr(VI) can be accomplished using many different *in situ* or *ex situ* methodologies. Table 1 presents a summary of the technologies discussed in this article. The selection of any particular technology requires the user to recognize the significant differences not only in the toxicological properties of Cr(VI) vs. Cr(III) but also in their physical/chemical behavioral differences under environmental conditions. *Ex situ* processes have been widely applied to chemically reduce Cr(VI) to Cr(III). These processes frequently include the addition of solidification agents to immobilize the trivalent Cr within the soil matrix. Many

TABLE 1
Treatment Methods for Cr(VI) in Soil

Method	Advantages	Disadvantages	Cost
Excavation and off-site disposal	Appropriate for small volumes of soil Quick to implement; uses readily available equipment Completely removes contaminants	May cause Cr(VI) to become airborne and a health hazard May invoke land disposal restrictions Can be expensive, especially for deep materials Soil may need treatment, adding to cost May necessitate removal of on-site structures or groundwater	$5 to $10 per ton for excavation, $100 to $200 per ton for disposal
Soil washing	May greatly reduce the volume of contaminated material that requires treatment	May cause Cr(VI) to become airborne and a health hazard during excavation Does not destroy contaminants Generates contaminated water and fines that require treatment Not appropriate for soils with >20% fines Not appropriate for soils with Cr(VI) in the soil particles	$50 to $200 per ton
Soil flushing	*In situ* technology, does not require excavation Demonstrated technology for remediating Cr(VI) spills Relatively inexpensive	Does not destroy contaminants Not appropriate for soils with Cr(VI) in the soil particles Generates contaminated water that requires treatment	$75 to $200 per ton
Solidification/ stabilization: *ex situ*	Demonstrated technology	May cause Cr(VI) to become airborne and a health hazard during excavation Does not destroy contaminants Generates increased volume of solidified mass that requires disposal Cr(VI) may first need to be reduced to Cr(III) Does not remove Cr; not appropriate for sites where total Cr levels must be decreased May require soils dewatering	$40 to $100 per ton

TABLE 1 (continued)
Treatment Methods for Cr(VI) in Soil

Method	Advantages	Disadvantages	Cost
Solidification/ stabilization: *in situ*	*In situ* technology, does not require excavation Applicable to sites with high water table	Cr(VI) may first need to be reduced to Cr(III) Does not remove Cr; not appropriate for sites where total Cr levels must be reduced	$10 to $30 for drilling and mixing alone; up to $200 per ton for remediation
Vitrification	May be performed *in situ* Should reduce and immobilize Cr(VI) May produce a marketable byproduct	High energy requirements May require soils dewatering May require significant amounts of additives	$350 to $400 per ton for *in situ* processes $350 to $700 for *ex situ* processes
Chemical reduction	May be performed *in situ* Applicable beneath buildings and other structures	Does not remove Cr; not appropriate for sites where total Cr levels must be decreased Other compounds in soil may consume reducing chemicals, and therefore process may require significant amounts of reducing agents	$75 to $100 per ton for *ex situ* process *In situ* process costs vary depending on application method
Biological reduction: *in situ*	*In situ* technology, does not require excavation Applicable to sites with high water table Stable reduction process, applicable to sites where Cr(VI) continues to leach from soil Applicable beneath buildings and other structures	Does not remove Cr; not appropriate for sites where total Cr levels must be decreased Requires control of pH, oxygen, nutrients, etc. Process may be slow to achieve Cr(VI) reduction	$20 to $100 per ton

types of reducing agents are available, ranging from various sulfite compounds to proprietary formulations and phosphate compounds. Ferrous sulfate is likely to be the reducing agent of choice for many soil matrices, due to its ability to work at neutral to alkaline conditions.

More recently, the treatment of Cr(VI)-bearing soils has focused on the application of *in situ* methods, such as chemical reduction coupled with stabilization/solidification or bioremediation. These processes typically use similar inorganic reagents as those used with *ex situ* methods or organic substrates containing bacteria to biochemically reduce the Cr(VI) to Cr(III). The potential promise of *in situ* technologies is the minimization of airborne particulates and the reduced cost of excavating and handling the soils and the possibility of treating both the Cr(VI)-contaminated soils and associated groundwater simultaneously.

Cleanup standards for Cr(VI)-contaminated soils traditionally have been based on the RCRA TCLP test for total Cr, even though most Cr contamination in soils is in the hexavalent form. With the trend of regulatory agencies toward the establishment of separate risk-based criteria for Cr(III) and Cr(VI), it appears likely that definitive soil cleanup criteria for each valence state will become commonly accepted in the near future.

The selection of a cost-effective remedial technology depends on determining the nature of the Cr(VI) present in the soils (i.e., surface contaminated or deeply embedded with the particles) and the other physical/chemical characteristics that may impact the treatment method performance, including the relevant site conditions, such as the depth to groundwater or bedrock or the presence of structures near or above the contamination.

For essentially all potentially applicable technologies, the performance of bench- or pilot-scale treatability studies is prudent to determine the process performance capabilities and unit costs for the site-specific soils. This is particularly important for innovative technologies such as bioremediation and electrokinetic processes, which are still in the development stage. Unit costs for Cr(VI) soil treatment range from \$40 to over \$400 per yd^3 for the various technologies that have been demonstrated at commercial scale or have been tested in pilot-scale sufficiently to make reasonable cost projections. The lower range applies to soils containing relatively soluble Cr(VI) that can be readily removed from the soil particles or can be stabilized/solidified in place without major complications associated with other contaminants or site conditions. The high end of the range applies to thermal treatment techniques, such as *in situ* vitrification, in which there are other contaminants of concern (e.g., organic and/or radioactive compounds) in addition to Cr(VI), which must be also treated.

REFERENCES

Armienta, M. A. and Quéré, A. 1995. Hydrogeochemical behavior of Cr in the unsaturated zone and in the aquifer of Leon Valley, Mexico. *Water Air Soil Pollut.* **84**, 11–29.

Arniella, E. F. and Blythe, L. J. 1990. Solidifying Traps. *Chem. Eng.* 92–102. February 1990.

Asp, J. 1990. Soil washing: a cost-effective remedial technology. *EI Digest.* October 1990, pp. 19–24.

Banerjee, S. J., Horng, J., Ferguson, J. F., and Nelson, P. O. 1987. *Field Scale Feasibility Study of Electrokinetic Remediation.* U.S. Environmental Protection Agency Risk Reduction Engineering Laboratory, Office of Research and Development, Cincinnati, OH.

Blake, R. C., Choate, D. M., Bardhan, S., Revis, N., and Barton, L. L. 1993. Chemical transformation of toxic metals by a *Pseudomonas* strain from a toxic waste site. *Environ. Geochem. Health,* **15(2),** 1365–1376.

Cabrera-Guzman, D., Swartzbaugh, J. T., and Weisman, J. T. 1990. The use of electrokinetics for hazardous waste site remediation. *J. Air Waste Manage. Assoc.* **40,** 1670–1676.

Conner, J. R. 1990. *Chemical Fixation and Solidification of Hazardous Wastes.* New York, Van Nostrand Reinhold.

Conner, J. R. 1994. Chemical stabilization of contaminated soils. In: *Hazardous Waste Site Soil Remediation: Theory and Application of Innovative Technologies,* pp. 128–130. (Wilson, D. J. and Clarke, A. N., Eds.) New York, Marcel Dekker.

DeFilippi, L. J. 1992. Bioremediation of hexavalent Cr(VI) contaminated solid residues using sulfate reducing bacteria. 4th USEPA Forum On Innovative Hazardous Waste Treatment Technologies: Domestic and International, San Francisco, CA, November.

Dragun, J. and Chiasson, A. 1991. *Elements in North American Soils.* Greenbelt, MD, Hazardous Control Resources Institute.

Environmental Science and Engineering, Inc. (ES&E). 1989. *Remedial Investigation for Chromium Sites In Hudson County, New Jersey.* Prepared for New Jersey Department of Environmental Protection.

Fisher, J. A. 1990. *The Chromium Program.* New York, Harper and Row.

Hanson, A. T., Dwyer, B., Samani, Z. A., and York, D. 1993. Remediation of Cr-contaminated soils by heap leaching: column study. *J. Environ. Eng.* **119(5),** 825–841.

Hazardous Waste Consultant (HWC). 1993. EPA Updates CERCLA Priority List of Hazardous Substances, **10(5),** 2.26–2.30, Lakewood, CO, McCoy and Associates, Inc.

Hazardous Waste Consultant. 1996a. Remediating Soil and Sediment Contaminated with Heavy Metals, **14(6),** pp. 4.1–4.57. New York, Elsevier Science.

Hazardous Waste Consultant. 1996b. Update on the SITE Demonstration Program, **14(1),** 1.1–1.11, New York, Elsevier Science.

James, B. R., Petura, J. C., Vitale, R. J. and Mussoline, G. R. 1995. Hexavalent chromium extraction from soils: a comparison of five methods. *Environ. Sci. Technol.* **29(9),** 2377–2381.

James, B. R. 1996. The challenge of remediating chromium-contaminated soil, *Environ. Sci. Technol.,* **30(6),** 248A–251A.

Lioy, P. J., Freeman, N. C. G., Wainman, T., Stern, A. H., Boesch, R., Howell, T., and Shupack, S. I. 1992. Microenvironmental analysis of residential exposure to chromium-laden wastes in and around New Jersey homes. *Risk Analysis,* **12(2),** 287–299.

Marvin, C. G. 1993. Government regulation increases for Cr-bearing hazardous waste. *Glass Industry* **74,** 16–17.

McPhillips, L. C., Pratt, R. C., and McKinley, W. S. 1991. Case history: effective groundwater remediation at the united chrome superfund site. Presented at the Hazardous Waste Management Assoc. Conference, Vancouver, British Columbia, June.

Mehlhorn, R. J., Buchanan, B. B., and Leighton, T. 1994. Bacterial chromate reduction and product characterization. *Emerging Technology for Bioremediation of Metals,* pp. 26–37. Boca Raton, FL, CRC Press.

New Jersey Department of Environmental Protection. 1990. Record of Decision for Hudson County Chromate Production Waste Sites: Operable Unit One-Residential Sites Remediation. April.

New Jersey Department of Environmental Protection. 1995. *Basis and Background for Soil Cleanup Criteria for Trivalent and Hexavalent Chromium.* Site Remediation Program.

Ohtake, H. and Hardoyo. 1992. New biological method for detoxification and removal of hexavalent Cr. *Water Sci. Technol.,* **25(11),** 395–402.

Oravetz, A. W., Smidt, S., Roth, E., and Davis, M. L. 1992. Variables that affect soil washing treatment for metals-contaminated soil. *Soil Remediation* 579–582.

Page, B. J. and Loar, G. W. 1991. Chromium compounds. In: *Kirk-Othmer Encyclopedia of Chemical Technology,* 4th ed., Vol. 6. (Kroschwitz, J. I. and Howe-Grant, M., Eds.) New York, Wiley-Interscience.

Patterson, J. W. 1985. *Hexavalent Chromium.* pp. 53–70. Industrial Wastewater Treatment Technology, 2nd ed., Boston, MA, Butterworth.

Richard, F. C. and Bourg, A. C. M. 1991. Aqueous geochemistry of Cr: a review. *Water Res.* **25(7),** 807–816.

Sheehan, P. J., Meyer, D. M., Sauer, M. M., and Paustenbach, D. J. 1991. Assessment of the human health risks posed by exposure to chromium-contaminated soils. *J. Toxic. Environ. Health* **32,** 161–201.

Shen, H. and Wang, Y-T. 1994. Biological reduction of Cr by *E. coli. J. Environ. Eng.* **120(3),** 560–572.

Shreve, R. N. 1967. *Chemical Process Industries.* New York, McGraw-Hill.

Stanek, J. 1977. *Electric Melting of Glass.* New York, Elsevier Science.

Stinson, M. K. 1990. EPA site demonstration of the international waste technologies/Geo-Con in situ stabilization/solidification process. *J. Air Waste Manage. Assoc.,* **40(11),** 1569–1576.

U.S. Department of Defense (USDOD). 1994. *Remediation Technologies Screening Matrix and Reference Guide.* Environmental Technology Transfer Committee. EPA/542/B–94/013, NTIS PB95–104782.

U.S. Environmental Protection Agency (USEPA). 1986. Stabilization/Solidification of Hazardous Waste. Office of Research and Development, Cincinnati, OH. EPA/600/D–86/028.

USEPA, 1988. Record of Decision for Odessa Cr I. Odessa, TX.

USEPA, 1989a. The Use of Electrokinetics for Hazardous Waste Site Remediation. Risk Reduction Engineering Laboratory, Cincinnati, OH.

USEPA, 1989b. SITE Program Demonstration Test International Waste Technologies In Situ Stabilization/Solidification, Hialeah, FL, Technology Evaluation Report, EPA Risk Reduction Evaluation Laboratory, Cincinnati, OH, EPA/540/5–89/004.

USEPA, 1990. Electrokinetics, Inc. The Superfund Innovative Technology Evaluation Program: Technology Profiles. EPA/540/5–90/006, pp. 129–130.

USEPA, 1992. The SITE Program: Spring Update to the Technology Profiles, 4th ed. Office of Solid Waste and Emergency Response. Office of Research and Development, Washington, DC. April, pp. 10–11.

USEPA, 1996a. Region III Risk-Based Concentration Table, January–June 1996, Memorandum from Roy L. Smith, Office of RCRA, Technical and Program Support Branch (3HW70).

USEPA, 1996b. Integrated Risk Information System (IRIS). Reference Dose for Oral Exposure to Cr (III) and Cr(VI). Cincinnati: Office of Health and Environmental Assessment, Environmental Criteria and Assessment Office.

USEPA, 1996c. Soil Screening Guidance: Technical Background Document. Office of Soild Waste and Emergency Response. EPA/540/R–95/128, Washington, DC., p. A–5.

Vitale, R. J., Mussoline, G. R., Petura, J. C., James, B. R. 1994. Hexavalent chromium extraction from soils: evaluation of an alkaline digestion method. *J. Environ. Qual.* **23(6),** 1249–1256.

Westbrook, J. H. 1991. Chromium and chromium alloys. In: *Kirk-Othmer Encyclopedia of Chemical Technology,* 4th ed., Vol. 6. (Kroschwitz, J. I. and Howe-Grant, M., Eds.) New York, Wiley-Interscience.

World Health Organization (WHO). 1988. *Chromium.* Environmental Health Criteria 61. Geneva, WHO.

CHROMIUM IN SOIL

Perspectives in Chemistry, Health, and Environmental Regulation

Edited by

Deborah M. Proctor
Brent L. Finley
Mark A. Harris
Dennis J. Paustenbach
and David Rabbe
CHEMRISK, A DIVISION OF MCLAREN/HART
PITTSBURGH, PENNSYLVANIA

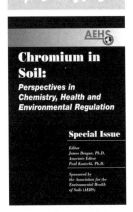

New!

AEHS

Chromium in Soil:
Perspectives in Chemistry, Health and Environmental Regulation

Special Issue

Editor
James Dragun, Ph.D.
Associate Editor
Paul Kostecki, Ph.D.

Sponsored by
the Association for the
Environmental Health
of Soils (AEHS)

As environmental standards are re-evaluated and modified, different issues take on different levels of importance. Today, the presence of chromium in soil and groundwater is one of the leading environmental issues. **Chromium in Soil** discusses the challenges faced by those investigating and remediating chromium-impacted soils and groundwater.

Features

- Coverage of the most recent research on chromium toxicity, chemistry, and environmental fate and transport
- New analytical methods from the EPA
- Results from human exposure studies
- Modern remediation technology and health-based cleanup standards
- Articles that have been written and reviewed by experts in the field

Catalog no. 1157, November 1997, 144 pp., ISBN: 0-8493-1157-8, $59.95

Mail, Fax, or Phone your order to:

LEWIS PUBLISHERS

2000 CORPORATE BLVD., N.W., BOCA RATON, FL 33431-9868

Tel: 1-800-272-7737 or Tel: 1-561-994-0555 for Florida and outside the continental U.S., or fax us toll-free **FAX: 1-800-374-3401**
Visit us on the **WorldWide Web: http://www.crcpress.com** Or **e-mail: orders@crcpress.com**
Prices quoted are in U.S. dollars and are subject to change

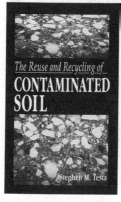

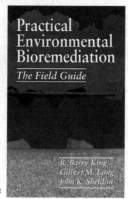

The nuts and bolts of site remediation today

SITE REMEDIATION

Planning and Management

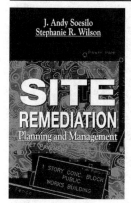

J. Andy Soesilo
HAZARDOUS WASTE SECTION, ARIZONA DEPARTMENT OF ENVIRONMENTAL QUALITY, AND CENTER FOR ENVIRONMENTAL STUDIES, ARIZONA STATE UNIVERSITY, TEMPE, ARIZONA

Stephanie R. Wilson
OFFICE OF CUSTOMER SERVICE AND EXTERNAL AFFAIRS, ARIZONA DEPARTMENT OF ENVIRONMENTAL QUALITY, PHOENIX, ARIZONA

Site Remediation: Planning and Management describes the management of remediation from a planning perspective, skillfully combining Superfund requirements and site remediation strategy in one practical volume. It clarifies and suggests remedies for the current quagmire of confusing Superfund reform and slow, expensive site remediation by thoroughly explaining the Superfund program and then describing how each of its components can fit into an integrated planning and management strategy. **Site Remediation** covers environmental sampling, site characterization, risk assessment, cleanup criteria, technology and technology screening, and public participation. Detailed and comprehensive, yet easy to understand, this book contains all you need to know about this important subject.

Features

- A comprehensive approach to site remediation that allows you to quickly grasp its nature and scope
- A summary of remediation regulations that eliminates confusion about which regulatory program a site must follow
- Detailed discussions of the site assessment planning process that helps you understand this complex process
- Thorough reviews of remediation technology that tell you what's available and how to choose effectively
- Identification of issues, problems, and trends that orient you to the current and future directions of site remediation

Catalog no. L1207, 1997, 432 pp., ISBN: 1-56670-207-0, $59.95

Mail, Fax, or Phone your order to:

LEWIS PUBLISHERS
2000 CORPORATE BLVD., N.W., BOCA RATON, FL 33431-9868
Tel: 1-800-272-7737 or Tel: 1-561-994-0555 for Florida and outside the continental U.S., or fax us toll-free FAX: 1-800-374-3401
Visit us on the WorldWide Web: http://www.crcpress.com Or e-mail: orders@crcpress.com
Prices quoted are in U.S. dollars and are subject to change